THE CAMBRIDGE GUIDE TO
THE
EARTH

THE CAMBRIDGE GUIDE TO

THE

EARTH

David Lambert
and the Diagram Group

The right of the
University of Cambridge
to print and sell
all manner of books
was granted by
Henry VIII in 1534.
The University has printed
and published continuously
since 1584.

CAMBRIDGE UNIVERSITY PRESS
Cambridge
New York   New Rochelle   Melbourne   Sydney

Acknowledgment
In the course of preparing this book for
publication we have consulted many
reference sources, including those listed on
page 248. If the publishers have unwittingly
infringed copyright in any illustration
reproduced they will gladly pay an
appropriate fee on being satisfied as to the
owner's title.

Published by the Press Syndicate of the University of Cambridge
The Pitt Building, Trumpington Street, Cambridge, CB2 1RP
32 East 57th Street, New York, NY 10022, USA
10 Stamford Road, Oakleigh, Melbourne, 3166, Australia

First published 1988

British Library cataloguing in publication data:

Lambert, David, *1932-*
    The Cambridge guide to the Earth.
    1. Earth
    I. Title   II. Diagram Group
    550   QE501

ISBN 0 521 33365 2 hard covers
ISBN 0 521 33643 0 paperback

Consultancy panel

FOREWORD

This book provides perhaps the most concise yet comprehensive key to the ingredients and processes that forged our planet. Hundreds of illustrations – large, clearly labeled diagrams, "field-guide" illustrations, and maps – help readers grasp important concepts instantly. Images are integrated with text explaining scientific terms in simple language. Together, text and pictures offer an up-to-date guide for all, from the inquiring eleven-year-old to the budding scientist.

There are twelve chapters. Each has a brief explanatory introduction, followed by topics arranged under bold headings.

Chapter 1 (Sizing up the Earth) gives an overview of our planet's origins and raw materials.

Chapter 2 (The Restless Crust) explains the astonishing processes that shape and reshape continents and oceans and recycle rocks.

Chapter 3 (Fiery Rocks) deals with igneous rocks – rocks formed from molten matter, the stuff from which all rock derives.

Chapter 4 (Rocks from Scraps) covers sedimentary rocks – rocks mostly formed from reconstituted bits of igneous or other rocks.

Chapter 5 (Deformed and Altered Rocks) investigates processes that displace crustal rocks and change one kind of rock into another.

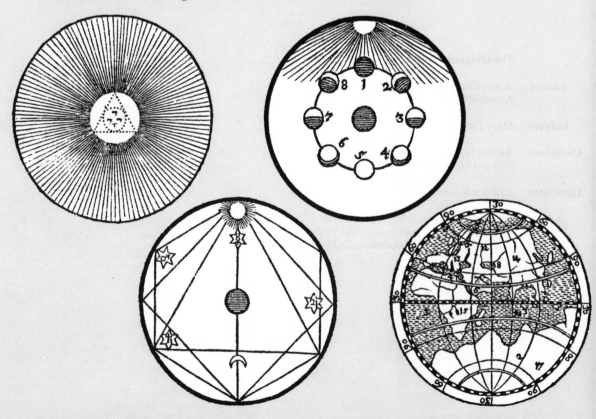

Chapter 6 (Crumbling Rocks) examines the work of weather in destroying rocks, building soil, and sculpting slopes.

Chapter 7 (How Rivers Shape the Land) traces the role of rivers in carving valleys, shaping plains, and deltas, and creating lakes.

Chapter 8 (The Work of the Sea) tells how the sea destroys and builds land.

Chapter 9 (The Work of Ice and Air) explains how glaciers and ice sheets mold cold lands, and winds sculpt desert landscapes.

Chapter 10 (Change Through the Ages) shows how geologists read the record of Earth's history in rocks, and what the oldest rocks reveal.

Chapter 11 (The Last 600 Million Years) covers the great changes that have formed, reformed, and repositioned continents since rocks acquired a detailed fossil record.

Chapter 12 (Rocks and Man) describes how geologists find and exploit useful rocks and minerals. The chapter ends with achievements of famous geologists and a worldwide list of geological displays.

Lastly there is a list of books for further reading, and an index.

The producers of this guide would like to thank the many experts whose work has helped to make this book more accurate and up to date.

CONTENTS

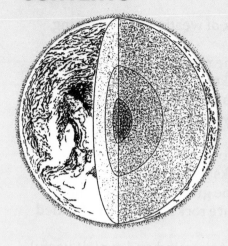

Chapter 4
ROCKS FROM SCRAPS

Chapter 5
DEFORMED AND ALTERED ROCKS

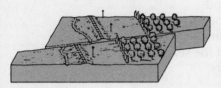

Chapter 6
CRUMBLING ROCKS

Chapter 7
HOW RIVERS SHAPE THE LAND

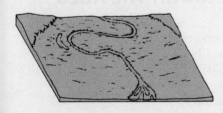

Chapter 8
THE WORK OF THE SEA

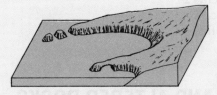

Chapter 9
THE WORK OF ICE AND AIR

Chapter 10
CHANGE THROUGH THE AGES

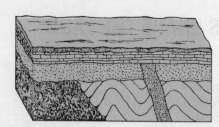

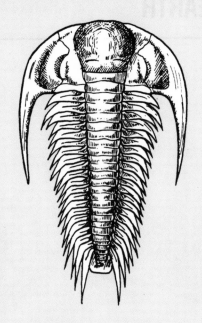

Chapter 11
THE LAST 600 MILLION YEARS

Chapter 12
ROCKS AND MAN

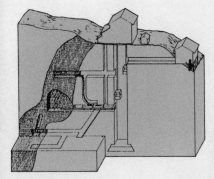

Chapter 1

SIZING UP THE EARTH

This book begins by putting our planet in its universal context. We see how matter, stars, and the solar system evolved, and how the Earth acquired its layered structure, and slightly bulging shape. There is a brief overview of elements, minerals, and rocks – Earth's building blocks. The chapter ends with a look at the forces that keep the sea and air in motion, disturb the crust, and turn the Earth into a mighty dynamo.

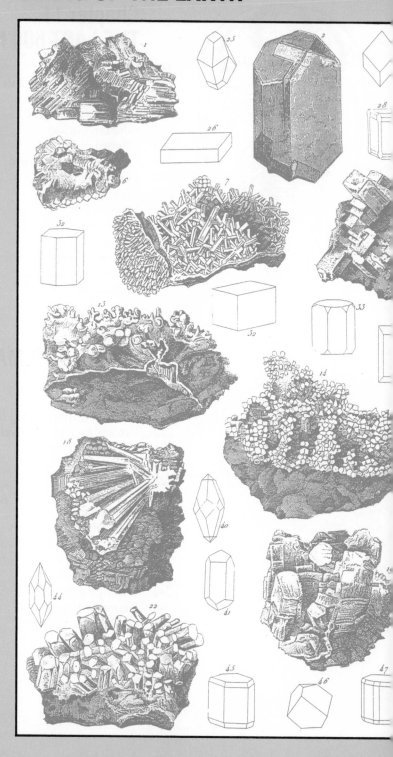

A selection of minerals and their crystalline forms. (Engraving originally published in *The Iconographic Encyclopaedia of Science, Literature and Art* 1851)

12

Earth in space

Earth is a rocky, spinning ball – one of nine planets and many lesser bodies (moons, asteroids, and comets) orbiting a star (the Sun). All of these together constitute our solar system.

The Earth is tiny compared to the four largest planets. But our solar system's largest and most influential body is the Sun, a glowing ball of gases a million times the volume of the Earth, and far bigger than all other objects orbiting the Sun. The Sun's immense gravitational force prevents these flying outward into space. And its electromagnetic radiations produce the heat and light that help to make life possible on Earth – the third nearest planet to the Sun. Most planets closer in or farther out appear too hot or cold for life.

Earth's behavior and that of its Moon determine time on Earth. Like all planets, Earth spins about a central axis with imaginary ends at the poles. Each rotation of about 24 hours produces day and night.

About once a month the Moon completes one revolution around the Earth. Earth itself completes one orbit of the Sun in about 365 days – an Earth year. Because Earth orbits in a tilted attitude, sunlight beams down directly upon its Northern and Southern hemispheres at different times of year, creating seasons.

Immense distances separate the Earth from other bodies in space. From Earth to the Moon is about 1.25 light-seconds, the distance covered in that time by light, which travels at 186,000mi per second (300,000km per sec). From Earth to Sun is 8 light-minutes; the solar system is 11 light-hours across; from Earth to the nearest star beyond the Sun is 4 light-years. Our solar system, plus dust, gas, and 100,000 million stars (some no doubt with solar systems of their own) comprise our galaxy – a flattened disk 80,000 light-years across. At least 10,000 million galaxies are scattered through the universe.

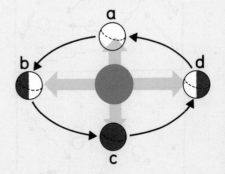

The seasons (above)
The Earth's tilt brings the midday sun overhead north or south of the equator at different times of year, so creating seasons.
a 21 March: Sun over the equator
b 21 June: Sun over the Tropic of Cancer
c 23 September: Sun over the equator
d 21 December: Sun over the Tropic of Capricorn

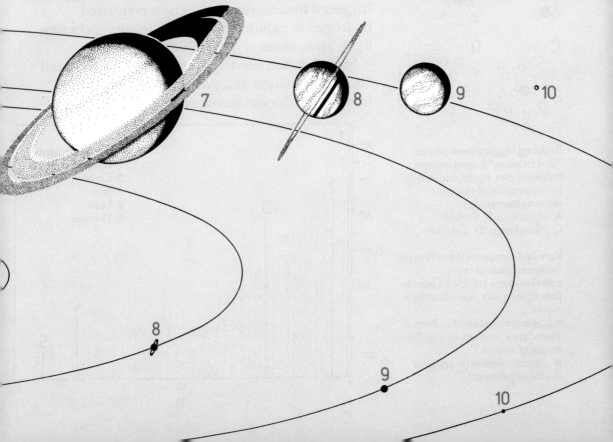

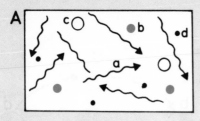

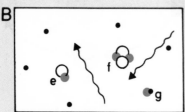

The first atoms (above)
A Post big-bang radiation (a) and
subatomic particles: protons (b),
neutrons (c), and electrons (d)
B Subatomic particles combined
as nuclei of deuterium ("heavy
hydrogen") (e), helium (f), and
hydrogen (g)

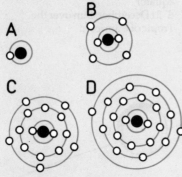

Building bigger atoms (above)
"Star factories" forged heavier,
more complex atoms, indicated
by the number of electrons
orbiting their nuclei:
A Hydrogen, B Carbon,
C Phosphorus, D Calcium

Rare and common elements (right)
Elements made of large,
complex atoms tend to be scarcer
than those made of small, simple
atoms.
a Logarithmic scale of universal
abundance (relative to 1 million
atoms of silicon.)
b Atomic number (number of
electrons per atom)

How everything began

Earth's origins lie in the creation of the universe. Just
how this came about remains unclear, but many
scientists accept some version of the "big bang"
theory, which goes like this. At first all energy and
matter (then only subatomic particles) was closely
concentrated. About 15,000 million years ago a vast
explosion scattered everything through space. Star
studies prove the universe is still expanding, and
background radiation hints at its initial heat.

The big bang sparked off processes producing
atoms forming different elements – the chemically
indivisible building blocks of stars and planets. (More
than 90 elements occur on Earth alone.)

Hydrogen and helium, the lightest, most abundant
elements, would have begun forming as subatomic
particles within minutes of the big bang. Hundreds of
millions of years later, condensing clouds of hydrogen
formed galaxies. Here mighty blobs of gas contracted
under gravitation. This process warmed the gas and
triggered nuclear reactions. These converted
hydrogen to helium and gave off energy including
light. Thus blobs of gas evolved into stars.

Stars that had used up all their hydrogen started
"burning" helium and swelling into "red giants."
Inside their nuclear furnaces atomic evolution

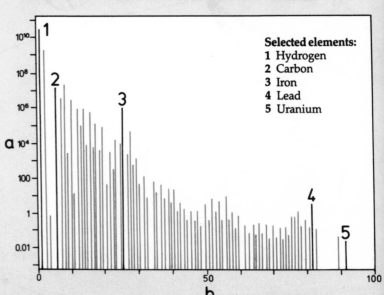

Selected elements:
1 Hydrogen
2 Carbon
3 Iron
4 Lead
5 Uranium

speeded up. First, helium nuclei combined to form carbon atoms. Next, carbon gave rise to heavier elements such as neon, nitrogen, and oxygen. Then came still heavier elements including iron. From red giants such substances escaped into space. Meanwhile massive "burned out" stars became supernovae, ending in immense explosions that forged such heavy elements as uranium and gold and hurled them into interstellar space.

New stars acquired these preformed elements. They make up only 1 per cent of the Sun's mass. But much higher percentages occur in planets like the Earth. The next pages explain how Sun and planets formed.

Life of a star

Many stars go through the stages pictured on this page.

1 Gases make up a formless mass called a nebula.

2 Gravitation makes the nebula contract to a regular shape.

3 Contraction heats the center of the nebula until nuclear reactions turn hydrogen to helium and it becomes a star.

4 The star completes its main contraction and starts a long, stable phase of energy output as a so-called yellow dwarf.

5 The star's hydrogen exhausted, its core shrinks, its surface expands and cools.

6 The star grows brighter and expands into a red giant.

7 The star burns helium, heats up, and grows enormously.

8 The expanding star may reach 400 times its former size.

9 Its nuclear energy used up, the star collapses to a tiny, dense, dim white dwarf.

10 The star ends as a black dwarf, no longer yielding heat or light.

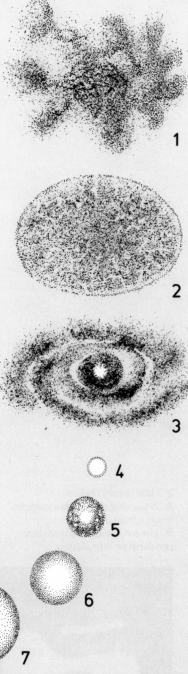

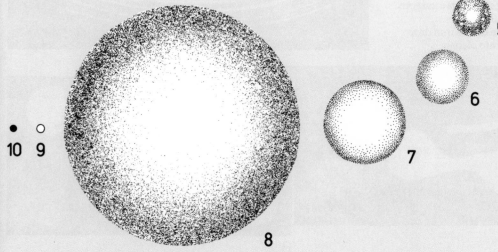

© DIAGRAM

Birth of the solar system

Cosmologists believe the universe was more than 10,000 million years old when our solar system formed 4600 million years ago. Its raw material was a cloud of dust and gas in one spiral arm of our galaxy. Somehow that cloud became the Sun and planets – a central star containing 99.8 per cent of the solar system and orbited by satellites lying roughly in one plane, most spinning in the same direction.

Scientists have produced various theories to account for this arrangement. Here are four. The fourth now seems most likely to be right.

1 Nebular hypothesis A spinning, shrinking cloud of gas and dust produced the Sun and threw off rings which condensed to form planets and their moons.

2 Tidal theory A passing star pulled a tongue of matter from the Sun. The tongue split into drops, then each drop became a planet.

1 Nebular hypothesis (right)
Planets form from rings of gas shed by the spinning center of a gas cloud.

2 Tidal theory (below)
A A passing star drags a gaseous tongue from the Sun.
B The tongue breaks into drops condensing into planets.

1

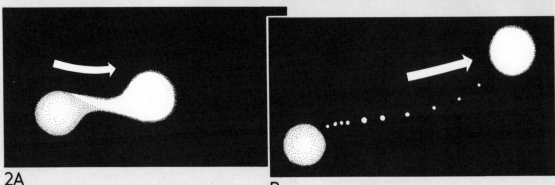

2A

B

3 Exploding supernova Our Sun supposedly had a companion star that exploded, leaving scattered debris that gave rise to planets.

4 Accretion theory Shock waves from supernovae made a cold cloud of gas and dust condense into smaller, denser clouds where complex molecules including methane formed. A collapsing portion of one cloud became a spinning disk of helium and hydrogen with 1-2 per cent of heavier elements. Gravitational attraction produced the Sun and the giant gas-rich planets. Meanwhile, dust grains stuck together and created lumps attracting smaller debris and growing into so-called planetesimals 60-600mi (100-1000km) across. After 100 million years or so, coalescing swarms of planetesimals formed the Earth and Earth-like planets – largely stripped of hydrogen and helium gases by a stream of solar particles.

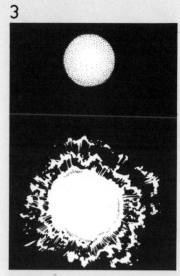

©DIAGRAM

3 Exploding supernova
The Sun's companion star explodes producing debris that forms planets.

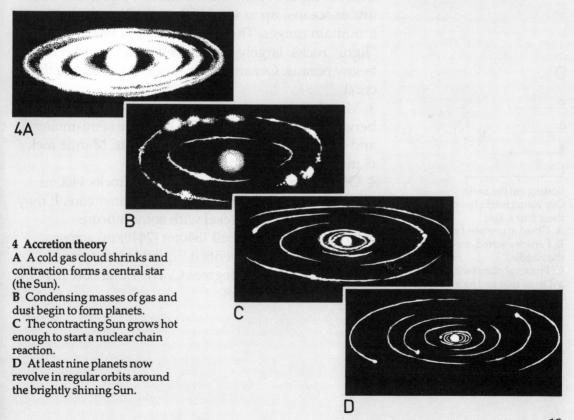

4 Accretion theory
A A cold gas cloud shrinks and contraction forms a central star (the Sun).
B Condensing masses of gas and dust begin to form planets.
C The contracting Sun grows hot enough to start a nuclear chain reaction.
D At least nine planets now revolve in regular orbits around the brightly shining Sun.

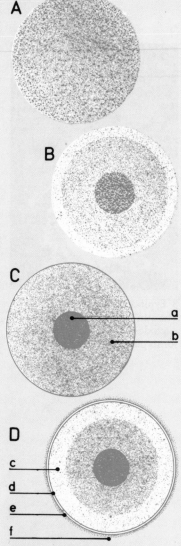

A

B

C

a
b

D

c
d
e
f

Sorting out the Earth
Our planet might have formed in these four stages.
A Cloud of unsorted particles
B Particles sorted, the densest in the middle.
C Primeval planet with:
a Dense iron and nickel core
b Material as in carbonaceous chondrite meteorites
D Outer chondrite melted, yielding:
c Relatively dense mantle
d Primitive crust
e Ocean
f Early atmosphere

Our layered planet

Accumulating mini-planets, dust, and gases formed the Earth about 4600 million years ago. Compression caused by gravitation produced immense internal heat and pressure. Meanwhile gravity was sorting out the Earth's ingredients. Heavy elements tended to be drawn into the middle. Lighter elements and compounds collected near the surface. The lightest elements included an outer film of gases.

This poisonous primeval atmosphere gave way to air in time, and Earth's molten surface cooled and hardened, but its interior remains intensely hot. (For more on the evolving Earth see Chapter 10.)

Such processes produced Earth's major layers:

1 Atmosphere An invisible film of gases (now mainly nitrogen and oxygen), from the base up divided into troposphere, stratosphere, mesosphere, ionosphere, thermosphere, exosphere, and magnetosphere.

2 Crust Earth's rocky outer surface – 4mi (6km) thick under oceans, up to 40mi (64km) thick under mountain ranges. The crust consists of relatively "light" rocks: largely granite below continents , basalt below oceans. Oceans cover about 71 per cent of the crust.

3 Mantle A layer of rocks 1800mi (2900km) thick between crust and outer core. Parts are semi-molten and evidently flow in sluggish currents. Mantle rock is more dense than crustal rock.

4 Outer core A layer of dense molten rocks 1400mi (2240km) thick between mantle and inner core. It may be mainly iron and nickel with some silicon.

5 Inner core A solid ball 1540mi (2440km) across. Intense pressure prevents it liquefying despite a temperature of 3700 degrees C. The inner core may be mainly iron and nickel.

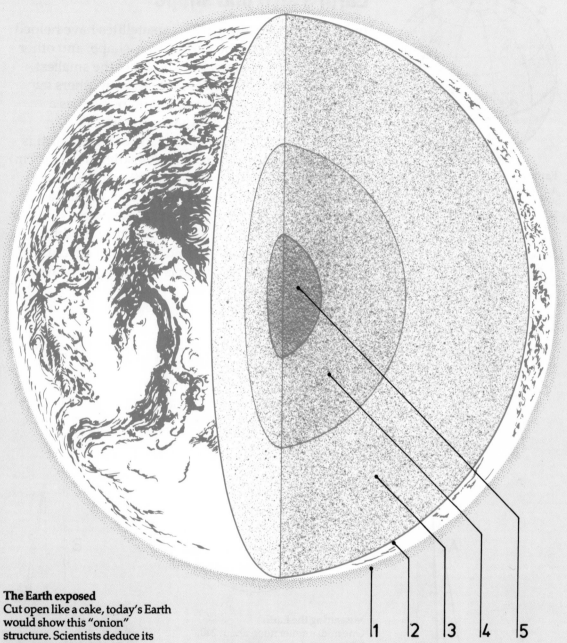

The Earth exposed
Cut open like a cake, today's Earth
would show this "onion"
structure. Scientists deduce its
inner layers from how these
interfere with waves set off by
earthquakes. (See also pp. 92-93.)
1 Atmosphere
2 Crust
3 Mantle
4 Outer core
5 Inner core

1　2　3　4　5

©DIAGRAM

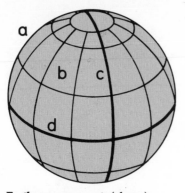

Earth measurements (above)
a Mass: 5976 million million million metric tons
b Surface area: 196,935 million sq mi (510,066 million sq km)
c Polar circumference: 24,859.7mi (40,008km)
d Equatorial circumference: 24,901.5mi (40,075km)

Earth's size and shape

Instruments including artificial satellites have helped scientists work out the Earth's size, shape, and other features. We now know ours is one of the smallest, lightest planets of the solar system. Four others far exceed its mass and volume. But no planet has a greater density (5.5 times that of water).

Careful measurements prove the ball-like Earth is not in fact a sphere. It measures 24,901.5mi (40,075km) around the equator, but only 24,859.7mi (40,008km) around the poles. So our planet bulges slightly at the equator and is slightly flattened at the poles. Centrifugal force created by Earth's spin produced this shape, an oblate spheroid. Even that description oversimplifies, for the Earth is very slightly pear-shaped.

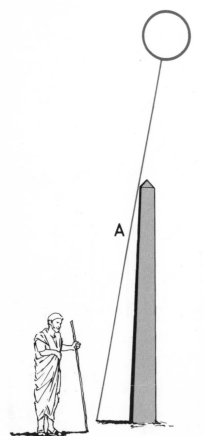

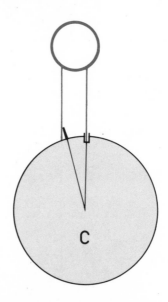

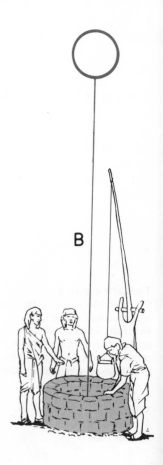

Measuring the Earth
One midsummer noon about 240 BC Eratosthenes used the Sun's parallel rays to find the Earth's circumference.
A Sunbeam in northern Egypt
B Sunbeam in southern Egypt
C The difference in angles let Eratosthenes work out the circumference.

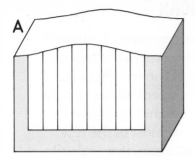
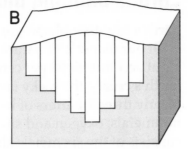

Scientists accordingly use the term geoid ("earth shaped") to describe the Earth's hypothetical, mean-sea-level surface – ignoring wrinkles formed by mountain chains and ocean floors. Geoid measurement involves taking sea-level gravity readings by gravimeter and studying "kinks" in orbits of artificial satellites. Both reveal so-called gravity anomalies reflecting local differences in mass in the Earth's crust and mantle. Such differences account for vast but slight dips and bumps in the geoid's surface.

Gravity anomalies also reinforce the theory of isostasy – a state of balance in the Earth's crust where continents of light material float on a denser substance into which deep continental "roots" project like the underwater mass of floating icebergs.

The global geoid (below)
This world map shows the global geoid's surface areas (a) above and (b) below the surface of a true ellipsoid, as deduced from surface gravity and artificial satellites. The geoid's highest "bumps" are little more than 170m (558ft) above its lowest depressions.

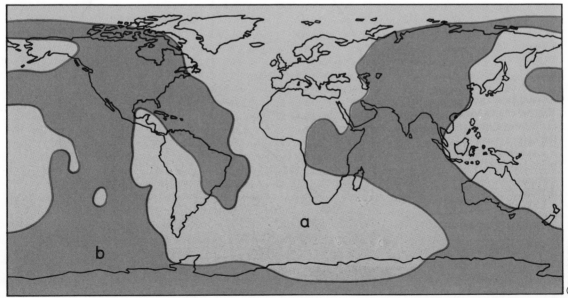

Earth's building blocks 1

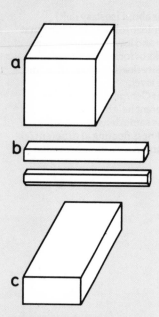

Crystal habits (above)
We show three types of habit –
crystal forms determined by
crystal system but modified by
factors such as temperature.
a Cubic
b Columnar
c Tabular

Earth's building blocks are elements and minerals.

Of our planet's 92 naturally occurring elements,
eight account for 98 per cent of the weight of the
Earth's crust – the rocky layer scientists know best.
Nearly three-quarters of its weight lies in two
nonmetals, oxygen and silicon; most of the rest
consists of the six metals aluminum, iron, calcium,
sodium, potassium, and magnesium.

Within the crust most elements occur as minerals –
natural substances that differ chemically and have
distinct atomic structures. The Earth's crust holds
about 2000 kinds of minerals. A few (including
gold) occur as just one element; most comprise two
or more elements chemically joined as compounds.
Silicates (minerals containing silicon and oxygen)
are the most abundant minerals in crust and mantle,
which make up four-fifths of our planet's volume.
(See also pp. 26-27.)

Most minerals formed from fluids that solidified – a
process that arranged their atoms geometrically,
producing crystals. Scientists identify six crystal

Crustal elements (right)
This pie chart shows percentages
of the most abundant elements in
the Earth's crust.
a Oxygen 46.6%
b Silicon 27.72%
c Aluminum 8.13%
d Iron 5.0%
e Calcium 3.63%
f Sodium 2.83%
g Potassium 2.59%
h Magnesium 2.09%
i Other elements 1.41%

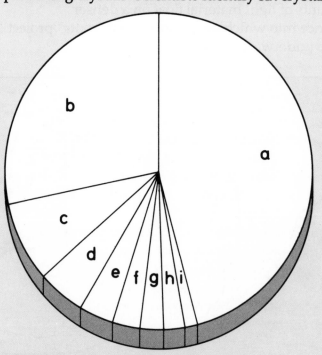

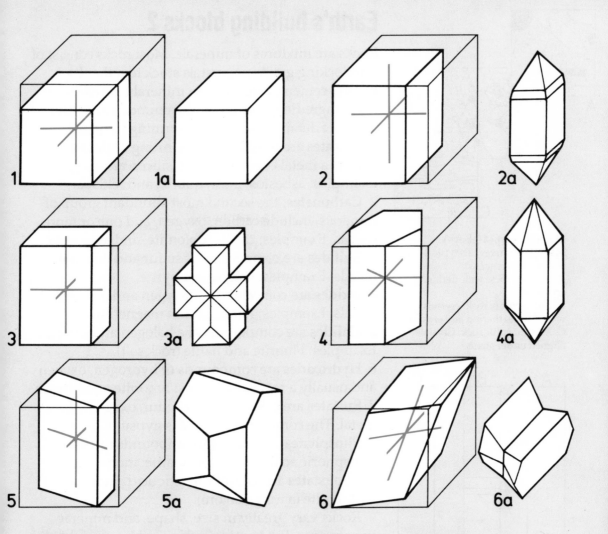

systems based on axes – imaginary lines passing through the middle of a crystal. Each system yields crystals with distinctive symmetry. Within each system, each mineral crystal grows in a special shape or habit, though this can be modified by temperature, pressure, and impurities.

Traditional tests identify minerals by hardness, color, streak (color of the powdered mineral), luster, specific gravity, cleavage, fracture, form, tenacity (resistance to bending, breaking, and other forces), odor, taste, and feel. Sophisticated tests involve polarizing microscopes, X rays, and spectral analysis.

Crystal systems
Systems vary with the number, lengths, and angles of axes. Internal axes are shown by colored lines.
1 Cubic
1a Example: halite
2 Tetragonal
2a Example: zircon
3 Orthorhombic
3a Example: staurolite
4 Hexagonal
4a Example: quartz
5 Monoclinic
5a Example: orthoclase
6 Triclinic
6a Example: albite

25

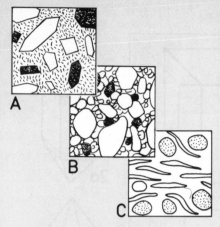

Earth's building blocks 2

Rocks are mixtures of minerals. Most rocks consist of interlocking grains or crystals stuck together by a natural cement. A few dozen minerals provide the main ingredients for the most common rocks. Here are brief details of some rock-forming minerals.

1 Silicates are the chief rock-forming minerals. Most feature a metal combined with silicon and oxygen. Examples: asbestos, mica, quartz, and feldspar.

2 Carbonates, the second most abundant group of minerals, include carbon, oxygen, and one or more metals. Examples: calcite, dolomite, and aragonite.

3 Sulfides are compounds of sulfur and one or more metals. Examples: galena and pyrite.

4 Oxides are compounds of oxygen and one or more metals. Examples: hematite and magnetite.

5 Halides are compounds of a halogen and a metal. Examples: Fluorite and halite (rock salt).

6 Hydroxides are compounds of hydrogen, oxygen and usually a metal. Examples: limonite and brucite.

7 Sulfates are compounds of sulfur, oxygen, and a metal. The commonest sulfate is gypsum.

8 Phosphates are chemical compounds related to phosphoric acid. Examples: apatite, monazite.

9 Tungstates are salts of tungstic acid. Example: wolframite (a tungsten ore).

Rocks vary greatly in size, shape, and mineral proportions. But texture (influenced by origin) helps geologists to identify three main groups: igneous, sedimentary, and metamorphic (see also Chapters 3-5).

A Igneous rocks have interlocking crystals formed as molten rock cooled down. The smaller the crystals the faster the cooling.

B Sedimentary rocks mainly have rather rounded mineral grains joined by natural cements. Most derive from the deposited remains of older rocks.

C Metamorphic rocks have crystals with a tendency to banding or alignment. They are formed from older rocks recrystallized by heat and pressure.

Major rock types (above)
Magnification reveals typical texture differences.
A Igneous rock: well-defined crystals
B Sedimentary rock: crystals worn by weathering and erosion
C Metamorphic rock: Crystals aligned under stress.

Mineral components
Bars show proportions of minerals in three rocks.
A Granite (igneous)
B Amphibolite (metamorphic)
C Shale (sedimentary)
a Quartz **b** Alkali feldspar
c Plagioclase feldspar
d Biotite **e** Magnetite
f Muscovite **g** Clay minerals
h Calcite **i** Amphibole

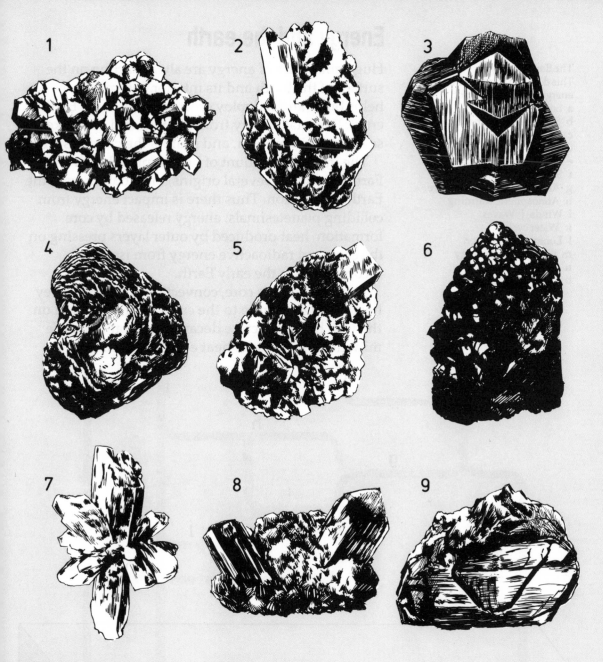

Rock-forming minerals
Illustrations above depict nine rock-forming minerals representing nine groups. (Note: one mineral can occur in several forms.) Numbers tally with those in the text.

1 Feldspar (a silicate)
2 Dolomite (a carbonate)
3 Pyrite (a sulfide)
4 Hematite (an oxide)
5 Fluorite (a flouride)
6 Limonite (a hydroxide)
7 Gypsum (a sulfate)
8 Apatite (a phosphate)
9 Wolframite (a tungstate)

Energy and the earth

Huge quantities of energy are always acting on the surface of the Earth and its interior. Internal heat helps build and redeploy great sections of the Earth's crust. External energy from the Sun and Moon keeps sea and air in motion, and sculpts the crustal surface.

The immense amount of heat trapped below the Earth's crust has several origins, most dating from the Earth's formation. Thus there is impact energy from colliding planetesimals, energy released by core formation, heat produced by outer layers pressing on the core, and radioactive energy from isotopes incorporated in the early Earth.

From the Earth's core, convection currents convey heat through mantle to the crust. More is added on the way by radioactive decay of crust and mantle minerals. Here, some heat escapes in violent

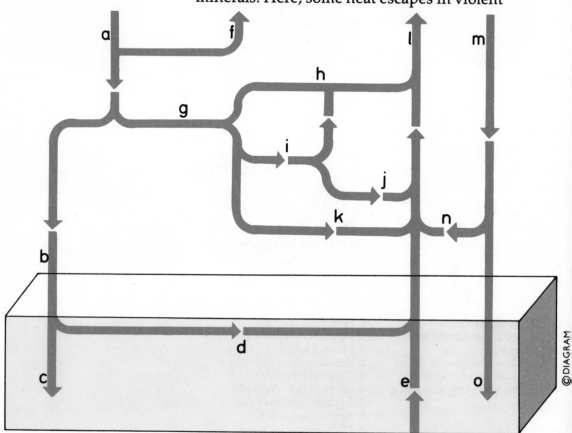

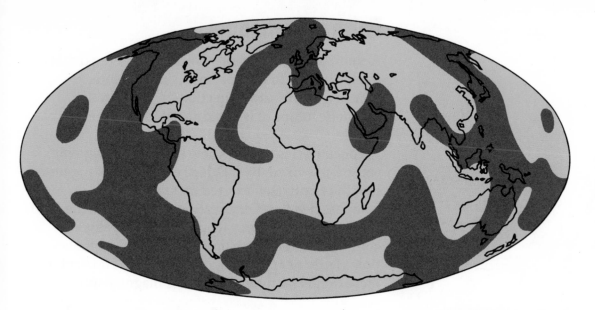

volcanic eruptions, hot springs, and earthquakes. But most simply leaks out quietly through continents and ocean floors – especially from the thin crust below some oceans.

The Earth's internal flow of heat produces most phenomena described in Chapters 2, 3, and 5.

The surface of the Earth's crust receives far more energy from above than below. Sunshine warms the atmosphere and crust. But sunshine warms tropics more than polar regions. This uneven heating creates belts of differing atmospheric pressure. Winds blow from high to low pressure areas. In turn, winds drive ocean waves and surface currents. Between them, winds and currents help to spread heat more evenly around the world. The Sun's heat also drives the water cycle. The resulting rain, rivers, glaciers, and ocean waves sculpt the surface of the land in ways described in Chapters 6 through 9.

Besides gaining heat from above and below, the surface of the Earth's crust loses heat by radiation into space. Loss roughly matches gain, so surface temperature has long remained about the same.

Lastly, the gravitational energy provided by the Moon and Sun produces our ocean tides, and tidal energy inside the Earth itself.

Heat-flow map (above) Strongest tones show areas of highest heat flow (largely ocean ridges) from the Earth's interior. Coolest areas include old parts of continents and ocean basins.

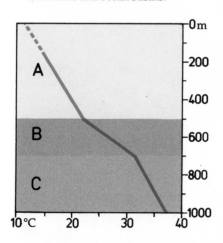

Temperature and depth
This graph shows temperature increasing from 59°F (15°C) near the surface to 100°F (38°C) at a depth of 1000m (3280ft). Rate of increase varies with the rock, as in our examples.
A Limestone
B Shale
C Granite

Earth as a magnet

About 2000 years ago the Chinese discovered that a freely-turning lodestone spoon always ended pointing in the same direction. Such naturally magnetic lumps of iron oxide led later to the compass needle.

For centuries no one knew just why a magnetic compass worked. By AD 1600 experiments suggested that the Earth itself exerted a magnetic force on compass needles. Observations proved that the Earth indeed has a magnetic field whose force aligns compass needles with north and south magnetic poles fairly near the geographic poles. (The field in fact extends far into space as the magnetosphere.)

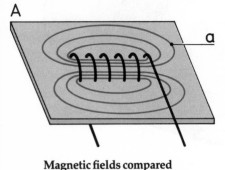

Magnetic fields compared
Above and right, two diagrams show the similarity between lines of force produced by an electric current flowing through a wire coil (**A**) and by the revolving Earth (**B**).
a Lines of magnetic force
b Mantle
c Core

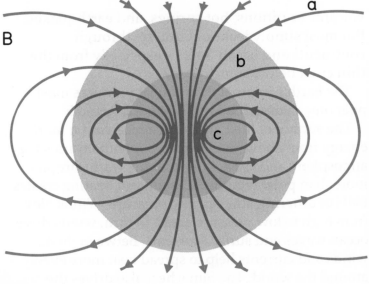

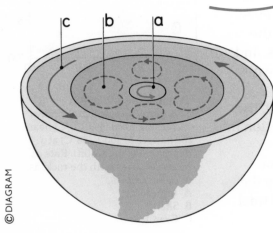

Inside the Earth (left)
The Earth cut open at the equator shows internal differences of rotation producing the magnetic field.
a Inner core rotation
b Eddies in the outer core
c Rotation of mantle

©DIAGRAM

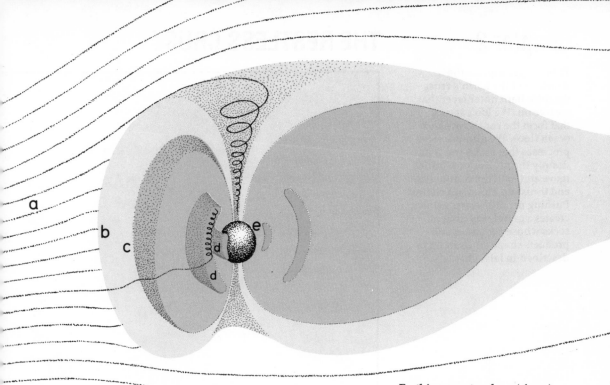

Just why the Earth should be magnetic became clear only in the middle 1900s. Scientists now realize that the Earth is less a huge bar magnet than a self-exciting dynamo. Inside the Earth radioactive heat keeps streams of molten metal flowing through the outer core. This process generates electric currents that produce magnetic fields – much as an electric current flowing through a coil of copper wire creates a magnetic field around the wire.

Earth's spin around its axis helps steer currents and create magnetic poles. Mighty eddies in the currents probably explain why magnetic poles slightly shift position from year to year. More puzzling are the hundreds of reversals of polarity occurring through Earth's history. Some rocks retain a record of the Earth's polarity at the time those rocks were formed. Paleomagnetism – the study of the Earth's magnetic field in prehistoric times – helps geologists to date certain rocks. It also helps them understand past and present movements of vast slabs of crust – the major subject of our second chapter.

Earth's magnetosphere (above)
a Solar "wind," streamlining Earth's magnetic field
b Shock front where solar wind meets magnetic field
c Magnetopause: edge of the magnetic field
d Van Allen radiation belts
e Earth

Polarity reversals (below)
This column shows epochs of predominantly normal (a) and reversed (b) polarity for the last 4.2 million years.
A Brunhes (normal)
B Matuyama (reversed)
C Gauss (normal)
D Gilbert (reversed)

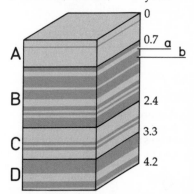

Chapter 2

THE RESTLESS CRUST

These pages explore the structure of the Earth's crust – the thin, hard outer layer that we live on. We look at rocks that form the continents and ocean floor. We glimpse slow processes that make and then destroy the ocean floor, build, move and split the continents, and thrust up mountain chains. Pushing up and wearing down creates a grand recycling of rocks whose processes and products shape lands in ways described in later chapters.

A representation of ocean-bed topography in which ocean ridges and underwater contours can be clearly seen.

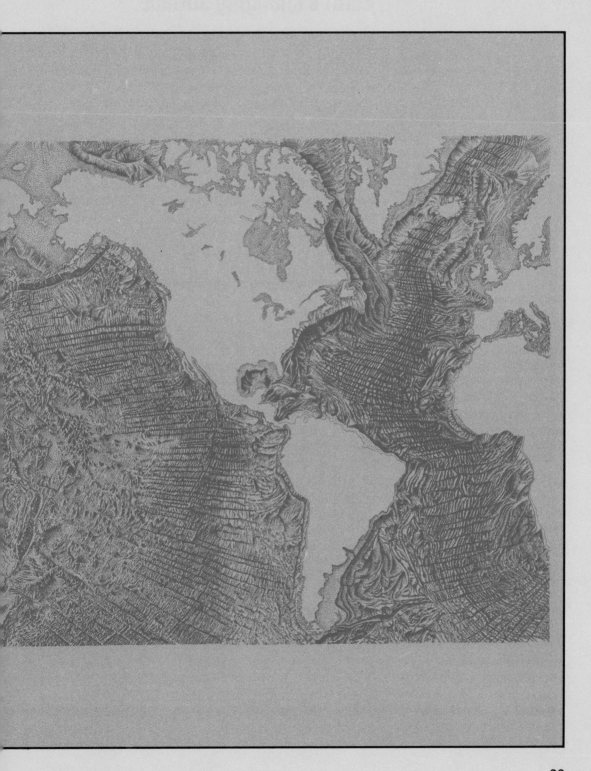

Earth's changing surface

Our planet's solid surface is a restless jigsaw of abutting, diverging, and colliding slabs called tectonic (or lithospheric) plates. How plates behave forms the subject known as plate tectonics.

Each plate involves a slab of oceanic crust, continental crust, or both, coupled to a slab of rigid upper mantle. Collectively, these plates make up the lithosphere. This rides upon the asthenosphere, a dense, plastic layer of the mantle. Heat rising through this layer from the Earth's core and lower mantle seemingly produces convection currents that shift the plates above.

Plate activities produce three main kinds of plate

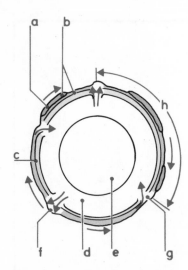

Plate tectonics in action (above)
This section through the world suggests that lithosphere is made by currents rising in the mantle at constructive margins, and lost at destructive margins where mantle currents sink.
a Continental crust
b Lithosphere
c Asthenosphere
d Lower mantle
e Core
f Constructive margin
g Destructive margin
h Lithospheric plate

Tectonic plates (right)
This map names major plates. Most plates are bounded by spreading ridges, and collision zones or subduction zones (marked by oceanic trenches). Active boundaries give rise to earthquakes or volcanoes.

⊔ Spreading ridges
— Collision zones
▲▲ Subduction zones
▨ Continental crust
⠤⠤ Volcanoes
⠂⠄⠂ Earthquakes

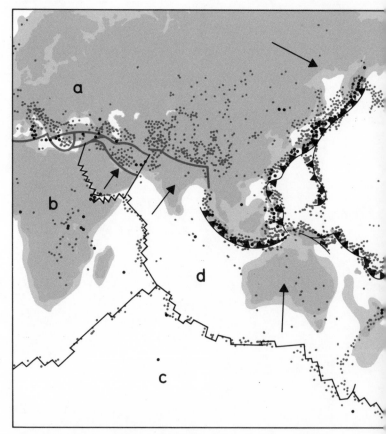

a Eurasian Plate   d Indo-Australian Plate   g Nazca Plate
b African Plate   e Pacific Plate   h South American Plate
c Antarctic Plate   f North American Plate

margins. Constructive margins are suboceanic spreading ridges where new lithosphere is formed between two separating oceanic plates. Destructive margins are oceanic trenches where an oceanic plate dives down below a (less dense) continental plate. Conservative margins are where two plates slide past each other and lithosphere is neither made nor lost.

Geophysicists also talk of active margins (where colliding continental and oceanic plates spark off volcanic eruptions, earthquakes, and mountain building) and passive margins (tectonically quiet boundaries between continental and oceanic crust).

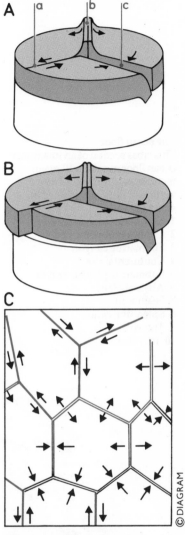

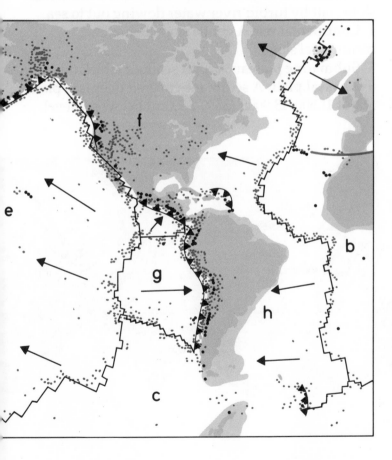

Plate boundaries (above)
AB Two block diagrams show three types of boundaries meeting as a triple junction between three tectonic plates.
A Plate positions now
B Plate positions later
a Conservative margins
b Constructive margins
c Destructive margin
C Bird's-eye view of possible boundary permutations where several plates interlock.

The ocean floor

Oceans and their seas hold 97 per cent of all surface water, and cover some 71 per cent of the Earth to an average depth of 12,400ft (3800m). Stripping off this watery sheath would reveal valleys, plateaus, peaks, and plains. We show ten features of the ocean floor.

1 Continental shelf A continent's true but submerged and gently sloping rim, descending to an average depth of 650ft (200m). Continental shelves occupy about 7.5 per cent of the ocean floor.

2 Continental slope A relatively steep slope descending from the continental slope. Such slopes occupy about 8.5 per cent of the ocean floor.

3 Submarine canyon A deep cleft in the continental slope, cut by turbid river water flowing out to sea.

4 Continental rise A gentle slope below the continental slope.

5 Submarine plateau A high seafloor tableland.

6 Abyssal plain A sediment-covered deep-sea plain about 11,500-18,000ft (3500-5500m) below sea level.

The ocean floor
This cross section of an imaginary ocean floor numbers 10 major features as in our text. (The vertical scale is exaggerated for effect.)
1 Continental shelf
2 Continental slope
3 Submarine canyon
4 Continental rise
5 Submarine plateau /guyot
6 Abyssal plain
7 Seamount
8 Spreading ridge
9 Trench
10 Island arc

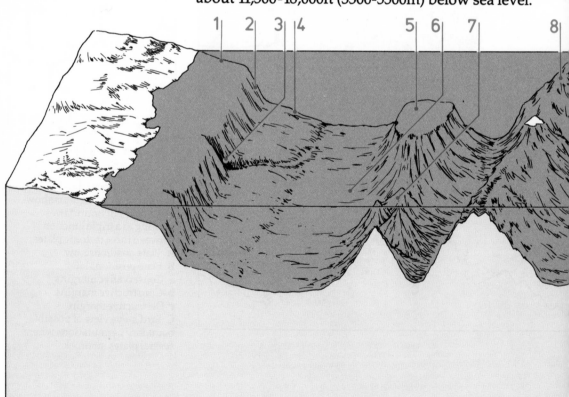

7 Seamount A submarine volcano 3300ft (1000m) or more above its surroundings. Guyots are flat-topped seamounts that were once volcanic islands.

8 Spreading ridge A submarine mountain chain generally 10,000ft (3000m) above the abyssal plain. A huge system of such ridges extends more than 37,000mi (60,000km) through the oceans. The Mid-Atlantic Ridge surfaces in places as volcanic islands such as Iceland and Ascension Island.

9 Trench A deep, steep-sided trough in an abyssal plain. At 35,840ft (10,924m) below sea level (deep enough to drown Mt. Everest), the Pacific Ocean's Marianas Trench is the deepest part of any ocean.

10 Island arc A curved row of volcanic islands, usually on the continental side of a trench.

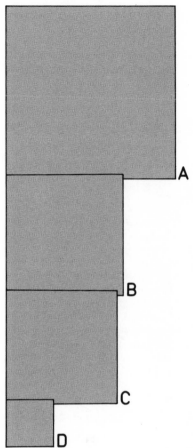

A

B

C

D

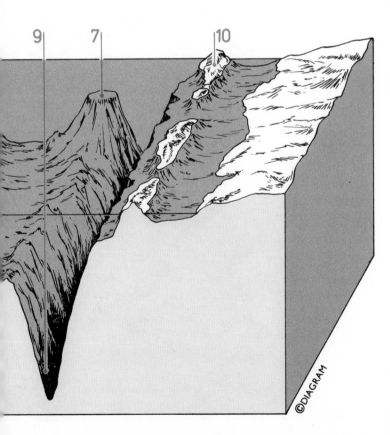

Ocean areas and depths
Diagrams contrast relative areas (above) and average depths (below) of the four oceans, omitting their marginal seas.
A Pacific Ocean
B Atlantic Ocean
C Indian Ocean
D Arctic Ocean

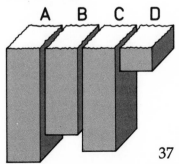

37

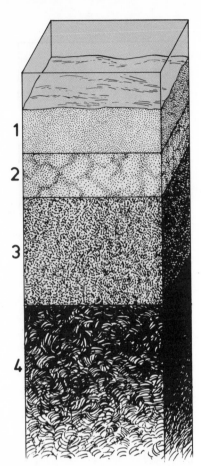

Oceanic crust

Dredging, boring, and seismic surveying suggest that oceanic crust is thinner, denser, and more simply made than continental crust. Oceanic crust is under 6.2mi (10km) thick. Its rocks are richer than mantle rocks in aluminum and calcium and their high silica and magnesium content earned oceanic crust the collective name of *sima*. Here are (simplified) details of oceanic crust's three layers, and an associated fourth layer.

1 Layer 1 The top layer consists of sediments. Muds, sands and other debris washed off continents lie up to ½mi (1km) thick on continental shelves and nearby ocean floor. The open ocean's bed bears oozes (remains of dead microorganisms from the surface waters), clays, and (in places) nodules rich in

The layered seabed (above) A block diagram shows one estimate of average depths of the four layers of oceanic lithosphere riding on the mantle. (Layer 1 thins away from continents.) Numbers match items in the text.
1 Layer 1
2 Layer 2
3 Layer 3
4 Layer 4

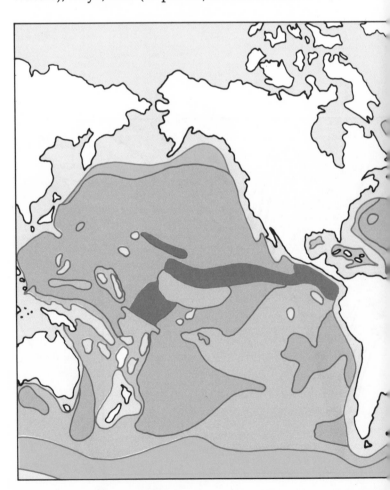

substances including manganese. No sediment occurs on spreading ridges.

2 Layer 2 is chiefly igneous rock, especially basalt, derived from the mantle and released at spreading ridges as rounded lumps of pillow lava. Scientists think the lower part of Layer 2 is seamed by sheeted dykes (see pp. 58-59). Much of Layer is 2 0.9-1.2mi (1.5-2km) thick.

3 Layer 3 is about 3mi (5km) thick and largely made of gabbro – a coarse-grained rock equivalent to the fine-grained basalt found in Layer 2.

4 Layer 4 is a rigid upper mantle layer coupled to the bottom of the ocean crust. It may be largely made of the dense igneous rock peridotite, consisting chiefly of the mineral olivine.

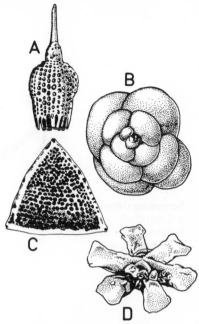

Sediment builders (above)
Four protists (enlarged) represent the marine microorganisms whose billions of limy or glassy skeletons form deep-sea sediments.
A Radiolarian (protozoan)
B Foraminiferan (protozoan)
C Diatom (protophyte with a siliceous cell wall)
D Coccoliths (calcareous plates from a protophyte)

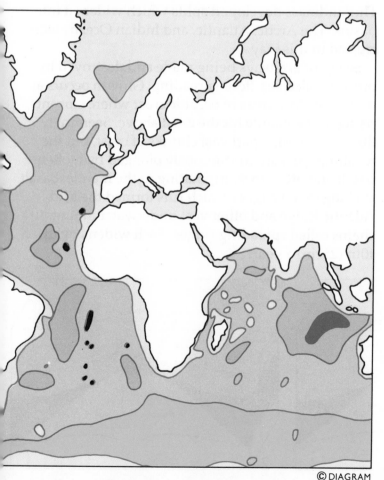

Ocean-floor sediments (left)
A world map shows the distribution of marine sediments, largely remains of land rocks, and oozes formed of tiny dead marine organisms.

- ☐ Terrigenous deposits (from eroded land rocks)
- ■ Red clay (from dust, etc)
- ☐ Foraminiferan ooze (calcareous)
- ☐ Pteropod (mollusk) ooze
- ☐ Diatom ooze (siliceous)
- ☐ Radiolarian ooze (siliceous)

© DIAGRAM

Sea-floor spreading

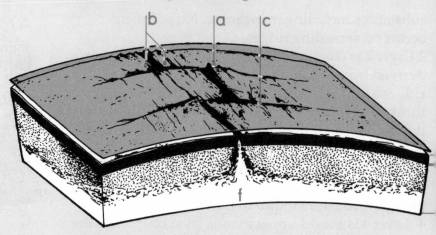

Spreading ridge (right)
a Central rift
b Spreading ridge
c Transform fault
d Oceanic crust
e Mantle
f Rising current

Spreading ridge evolving (below)
Diagrams show how a spreading ridge grows. (See page 41 for labels.)
1 A rising mantle current partly melts and pushes up the oceanic crust above.
2 The bulging plate splits and a central block subsides. Molten rock rises through cracks at the block's rims.
3 More molten rock plugs gaps left as tension pulls old crust apart.
4 As new crustal blocks subside, fresh cracks appear. The process is repeated.

Scarcely any ocean floor is more than 200 million years old. That long ago a single mighty ocean incorporating the Pacific surrounded one landmass. The landmass developed splits which widened into basins. The Arctic, Atlantic, and Indian Oceans were created in this way.

Sea floor is always being made and destroyed by a process called sea-floor spreading. Growth occurs at high heat-flow areas of oceanic crust where currents rising in the mantle hit the crust above. Seemingly this helps to tug apart vast chunks of crust, but the resulting gaps are continuously plugged by molten basalt and other rock originating in the mantle. Basalt sticking to the edges of such rifts formed the Mid Atlantic Ridge and other vast underwater mountain chains called spreading ridges. Each widens by up to 10in (25cm) a year.

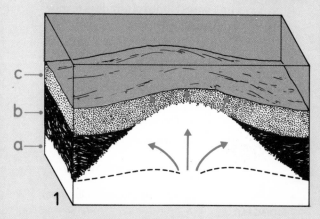

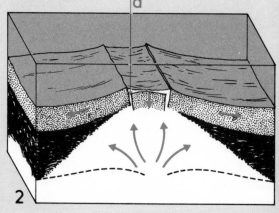

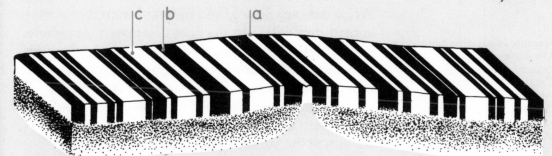

Scientists believe the detailed process goes like this. First, molten rock wells up from deep down in the upper mantle region called the asthenosphere. This great convection current partly melts the rocks around it to make the oceanic crust. Gravity pulls the ridge flanks down and sideways. The resulting tension opens two main cracks along the ridge. Between these cracks the ridge's middle sinks to form a central rift valley. Molten rock wells up through main and lesser cracks, then cools and hardens to become new ocean floor. Injections of fresh molten rock keep this spreading outwards from the central rift.

As upwelling continues, the rifting process is repeated. In time, rows of parallel ridges creep outward from their starting point, gradually sinking down to form the ocean's abyssal plains.

Meanwhile great cracks called transform faults cut across the central ridge at right angles, offsetting short, straight sections.

Fossil magnetization pattern
a Central rift
b Normally magnetized basalt
c Reversely magnetized basalt
As upwelling basalt forms new sea floor it takes on the polarity of the magnetic field at that time. So a spreading sea floor records geomagnetic reversals occurring every few hundred thousand years.

Labels for diagrams (below)
a Asthenosphere (part melted)
b Peridotite upper mantle, formed from dense minerals
c Gabbroic lower crust, from lighter minerals solidifying as lower ocean crust
d Basalt and dolerite – fine-grained gabbroic rocks formed by fast cooling

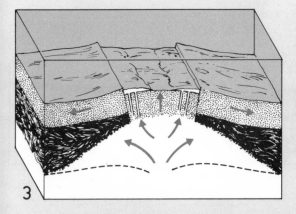

3

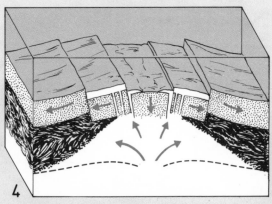

4

©DIAGRAM

How sea floor disappears

While new sea floor grows outward from the spreading ridges, old sea floor disappears elsewhere, subducted – drawn down – into the mantle. Oceanic trenches are the sites of these subduction zones where leading edges of lithospheric plates plunge under less dense or less mobile plates and so below a continent or ocean floor. Most trenches lie around the rim of the Pacific Ocean. Here vanish rocks originating from the spreading ridge that runs from Canada to south of New Zealand.

Subducted oceanic crust injects a tongue of relatively cool material into the hot mantle rock beneath. The friction of its passage generates earthquakes.

1 Oceanic subduction zone
a Oceanic trench
b Volcanic island arc
c Crust
d Mantle
e Area where oceanic crust is destroyed

2 Subduction zone beneath the South American continent
a Peru-Chile trench
b Sea floor (oceanic crust)
c Western cordillera (Andes)
d Eastern cordillera (Andes)
e Active volcano
f Continental crust
g Mantle

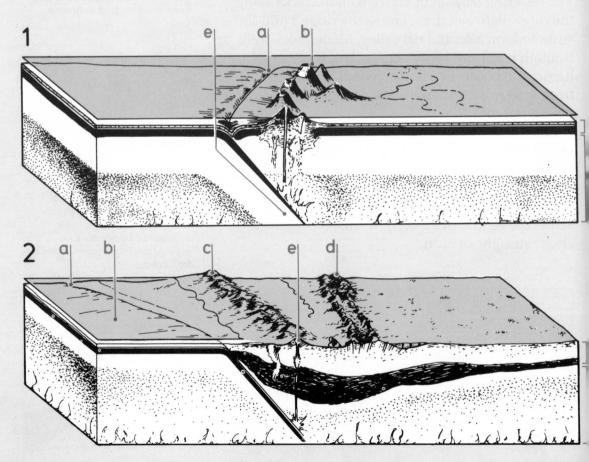

As the subducted oceanic crust descends, its load of low-density sediments is largely scraped off and deformed. Meanwhile, 60mi (100km) down, the sinking lithosphere begins to melt; and 430mi (700km) down it has completely broken up.

Less dense than the surrounding mantle, molten matter from subducted lithosphere bobs up again. This molten rock melts holes through the edge of the plate above the one subducted. Together with the scraped-off sediments, this process builds island arcs – rows of volcanic islands curved because they form upon the Earth's curved surface. (Pressing a ping-pong ball with your thumb similarly forms an arc-shaped fold.) Examples of such arcs occur in the Aleutian, Japanese, Kurile, and Cycladic Islands.

Oceanic crust subducted below a continental rim throws up volcanoes on the mainland. Volcanoes formed this way crown the Andes mountain chain of western South America.

Spreading and subduction
One diagram shows the complete conveyor-belt sequence of ocean-floor production, transportation, and destruction.
a Spreading ridge
b Transform fault
c Subduction zone
d Oceanic crust
e Continental crust

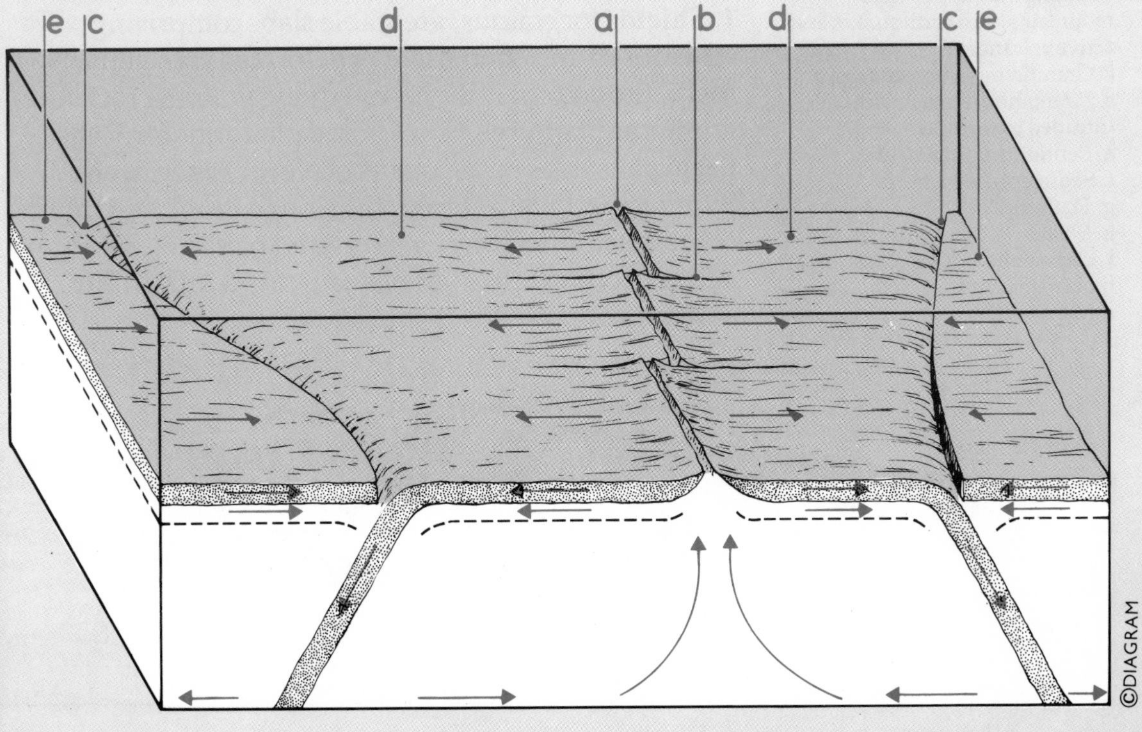

©DIAGRAM

43

The continental crust

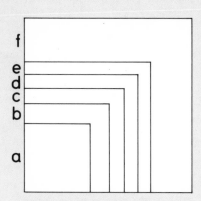

The continents (above)
Relative areas of continental lands
above sea level:
a Australia **b** Antarctica
c South America
d North America
e Africa **f** Eurasia

Inside a continent (below)
Section across an imaginary
landmass:
a Continental shelf
b Young mobile belt (fold
mountains with earthquakes and
active volcanoes)
c Granitic/metamorphic rocks
d Granodioritic crust with
intruded basic rocks
e Peridotite upper mantle
f Sedimentary basin
g Platform
h Shield
i Old mobile belt (old fold
mountains without earthquakes
or volcanoes)

Continents are the great land masses above the level of the ocean basins. The six major masses are North America, South America, Eurasia, Africa, Australia, and Antarctica. With their submerged offshore continental shelves these form 29 per cent of the Earth's surface, and 0.3 per cent of the Earth.

Continents are thicker, less dense, and contain more complex rocks than ocean crust. Continental crust averages 20mi (33km) thick, but can be twice that deep below high mountains. The top 9mi (15km) or so consists of sedimentary, igneous, and metamorphic rocks rich in silicon and aluminum, hence the collective name *sial* often used for continental crust. The lower crust has denser igneous and metamorphic rocks. Continental lithospheric plates ride on the asthenosphere.

Continents average 2950ft (900m) above sea level, but have wrinkles in the form of mountains, valleys, plains, and plateaus. Geologists consider continents contain these three main structural components:

1 Shields (or cratons) are stable slabs comprising outcrops of ancient masses of deformed crystalline rocks. Eroded shields plus overlying younger rocks are known as platforms. Shields and platforms form vast flat expanses, as in the plains of North America and Siberia, the Sahara Desert, the Congo Basin, and much of the Australian interior. Upraised parts of shields and platforms form high plateaus, especially in Africa and Asia.

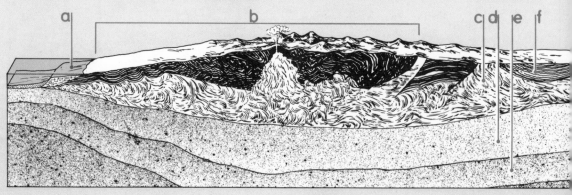

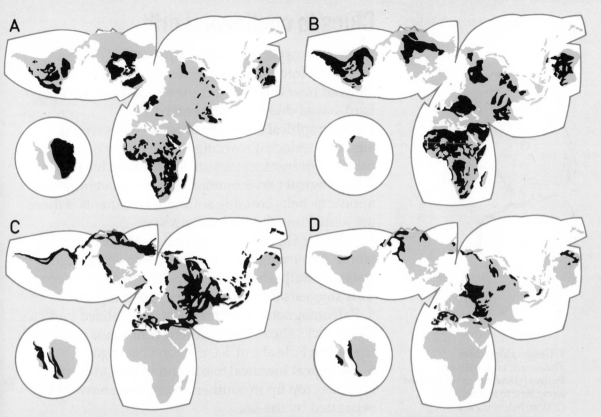

A

B

C

D

2 Linear mobile belts include young fold mountains like the Alps, Himalayas, Andes, and North American Cordillera. (See pp. 50-51.) Beveled, mobile belts probably comprise the long rock structures seen around the rims of ancient shields.

3 Sedimentary basins are broad, deep, depressions filled with sedimentary rocks formed in shallow seas that sometimes covered parts of ancient shields or their flanking mobile belts.

Continents' components (above)
These maps reveal continents as ancient cores with other units added later.
A Ancient rocks
B Sediment covering **A**
C Old and young fold mountains
D Sediment covering **C**

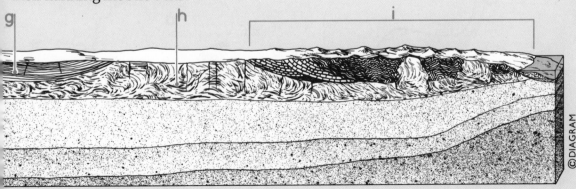

g   h   i

Clues to continental drift

Scientists suspected that continents had moved around before the discovery of sea-floor spreading showed how this might come about. Here are five land-based clues to continental drift.

1 Geographical Some continents' coasts would almost interlock if rearranged like pieces of a jigsaw puzzle. For instance, South America fits into Africa.

2 Geological Old mountain zones of matching ages appear as belts crossing southern continents if these are joined together in a certain way.

3 Climatic Glacial deposits and rocks scratched by stones in moving ice show that ice covered huge tracts of southern continents 300 million years ago. This suggests these places lay in polar regions.

4 Paleomagnetic Alignments of magnetized particles in old rocks show that southern continents all lay near the South Pole about 300 million years ago.

5 Biological Identical fossil land plants and land animals crop up in southern continents now widely separated by the sea.

1 Geographical clues
The Americas (**a, b**) fit into Europe (**c**) and Africa (**d**) if joined along their true rims, 6600ft (2000m) below the sea.

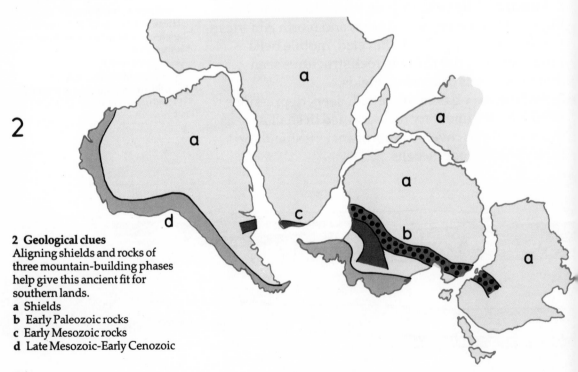

2 Geological clues
Aligning shields and rocks of three mountain-building phases help give this ancient fit for southern lands.
a Shields
b Early Paleozoic rocks
c Early Mesozoic rocks
d Late Mesozoic-Early Cenozoic

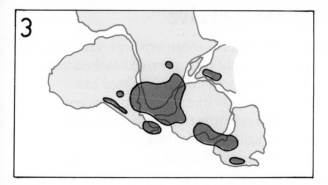

3 Climatic clues
About 320 million years ago south polar ice sheets could have straddled southern lands like this.

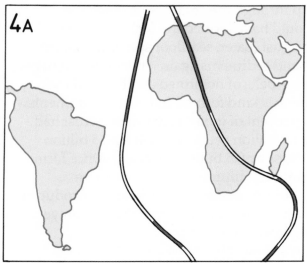

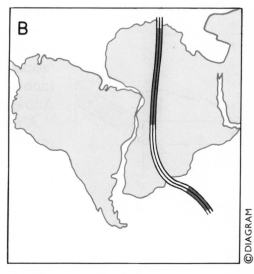

4 Paleomagnetic evidence
A Apparent wander paths of South America and Africa in relation to magnetic poles 400-200 million years ago.
B Paths coincided if both continents were joined.

5 Biological clues
These three fossil land organisms crop up in several southern lands as shown.
a *Glossopteris* (plant)
b *Lystrosaurus* (reptile)
c *Mesosaurus* (reptile)

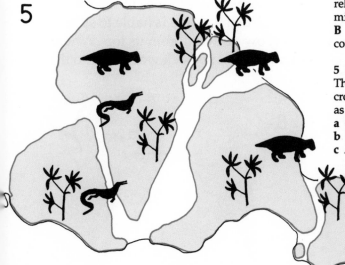

How continents evolve

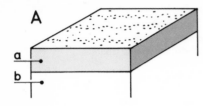

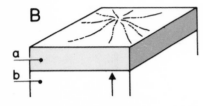

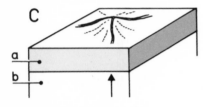

A rifting continent (above)
A Continental crust intact
B Crust bulges above a rising hot spot in the mantle
C Crust fractures
a Crust **b** Mantle

Triple junctions (below)
A A triple junction split South America from Africa.
B A triple junction splits Arabia from Africa. The Benue Trough (**a**) and Afar Depression (**b**) are both aulacogens.

Close study of the rocks of continents reveals ancient cores with progressively younger rocks tacked on to their rims. Each core, or craton, originated as a microcontinent, possibly like this. Two converging, cooling, horizontal currents in the mantle tugged on a tract of thin, early crustal rock, then sank. This squashed and thickened that patch of crust. Its base bulged down and melted, releasing light material that punched up through the crust above. Such rock resorting could have formed the first small slabs of continental crust. Later, sea-floor spreading swept island arcs and sediments against microcontinents as mobile belts – belts of deformed and buckled rock. Accretion of this kind formed full-blown continents.

About 5 per cent of today's continental crust had formed by 3.5 billion years ago, half by 2.5 billion years ago, most by 0.5 billion years ago. Once formed, continents are not immutable – they can be reworked, but not destroyed. Coalescing produced the supercontinent Pangea about 300 million years ago. Rifting later broke it up.

The Earth's crust splits open above "hot spots" – fixed plumes of molten rock rising in the mantle. A plume formed the volcanic Hawaiian Islands by punching through the thin oceanic Pacific Plate passing over it. Plumes raise domes in the thick, rigid continental crust. A dome is liable to split in three as cracks grow outward from its top. Where three cracks widen, oceanic rock wells up into the

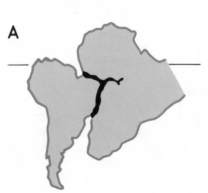

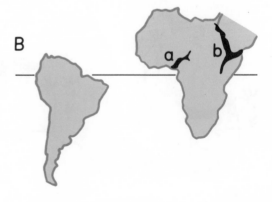

spreading gaps. The continent is split apart, and a triple junction then separates three lithospheric plates. If spreading happens only in two cracks, two plates form. The third crack becomes an abandoned trough or rift. Nigeria's Benue Trough and Ethiopia's Afar Depression are two such so-called aulacogens.

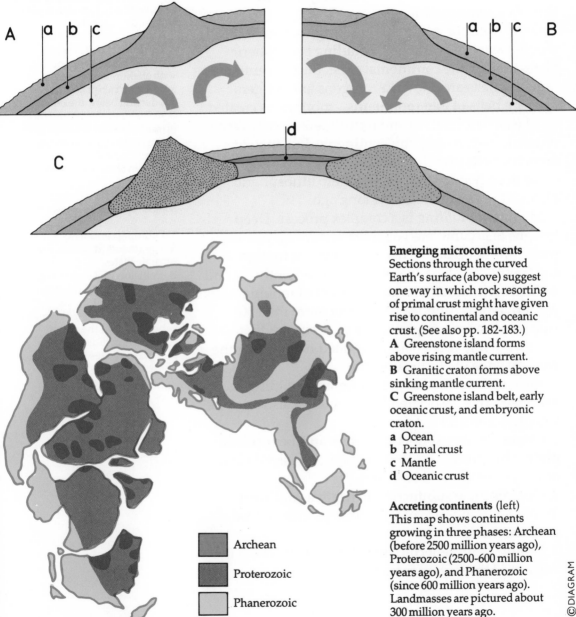

Emerging microcontinents
Sections through the curved Earth's surface (above) suggest one way in which rock resorting of primal crust might have given rise to continental and oceanic crust. (See also pp. 182-183.)
A Greenstone island forms above rising mantle current.
B Granitic craton forms above sinking mantle current.
C Greenstone island belt, early oceanic crust, and embryonic craton.
a Ocean
b Primal crust
c Mantle
d Oceanic crust

Accreting continents (left)
This map shows continents growing in three phases: Archean (before 2500 million years ago), Proterozoic (2500-600 million years ago), and Phanerozoic (since 600 million years ago). Landmasses are pictured about 300 million years ago.

Archean

Proterozoic

Phanerozoic

© DIAGRAM

49

Mountain building

Large regions of the Earth consist of mountains. Most occur in rows called ranges. Parallel ranges and intervening plateaus form chains such as the Andes and North American Cordillera. Related mountain chains and ranges make up mountain systems, notably the Tethyan (Alpine-Himalayan) and Circum-Pacific systems.

Orogenesis, or mountain building, occurs along mobile belts – places where colliding lithospheric plates disrupt the continental crust. Such mountain-building belts are known as orogens and orogenic belts are belts of fold mountains – mountains created by crustal deformation and uplift. Geologically recent orogenic belts mostly rim continents. But ancient orogenic belts (the Ural Mountains for example) can occur deep inside a continent where lithospheric plates were welded together long ago.

Mountain building is a complex process. Deep troughs of accumulated offshore sediment, volcanic rocks, bits of oceanic crust, and scraps of foreign continents can all be swept against one continent and welded on as mountain ranges. Most of mountainous western North America consists of more than 50 suspect terranes – mighty slabs of alien rock that independently rotated and migrated north along the western edge of North America.

We illustrate three major mountain-building processes. (For associated landforms see also Chapters 3 and 5.)

1 Oceanic plate subduction below another oceanic plate. This process created the Aleutian Islands and other mountainous island arcs.

2 Oceanic plate subduction beneath a continent. Involving island-arc collision, this process helped produce the Andes.

3 Double continent collision. The way in which the Alps and Himalayas formed.

1 Island-arc orogeny
a Subducted oceanic crust
b Low outer island-arc of sediments squeezed by subducting oceanic crust.
c Inner island-arc of mountainous volcanoes, produced by the "bobbing up" of light, subducted, melted oceanic crust and sediments.

2 Cordilleran belt orogeny
A Island-arc (c) and continent (d) with offshore sediments (b) advance on two plates, one subducting below the other.
B Collision squeezes and rucks up sediments (b) between island-arc volcanoes (c) and continent (d), producing a cordilleran mountain chain such as the Andes.
C and D The old subduction zone is replaced by a new one.

3 Colliding continents
A Continents (d) advance on separate plates.
B Collision rucks up marginal sediments (b) and the ocean shrinks.
C The oceanic crust between is subducted and the two continents collide, forming mountain ranges like the Alps or the Himalayas

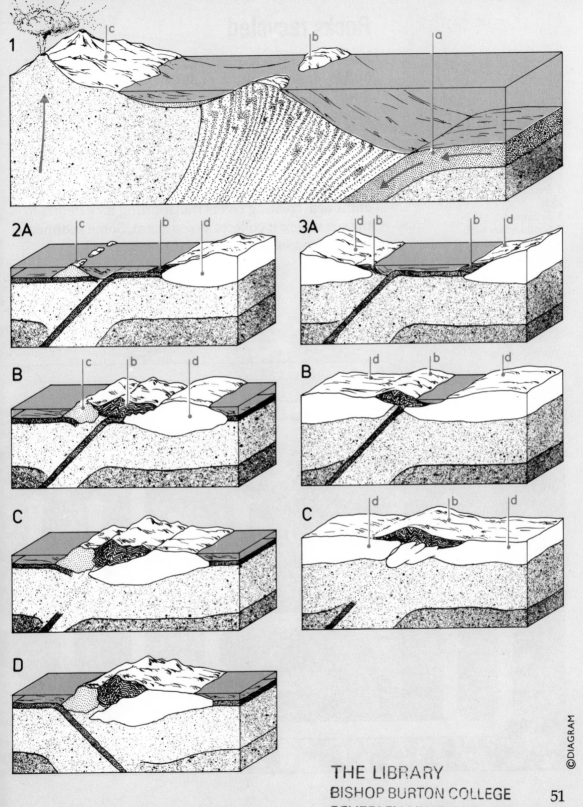

51

©DIAGRAM

Rocks recycled

Internal and external forces produce a rock cycle that builds, destroys, and remakes much of the rock that forms our planet's crust. The internal forces are produced by currents of rock that rise and spread out in the mantle, so moving lithospheric plates about. The main external forces are the weather's, generated by the energy in sunshine.

Weather wears down rocks exposed above the level of the sea, creating rivers transporting rock debris to the sea where it collects as sediment. Some sediments become consolidated into sedimentary rock.

Materials
1 Molten rock in the mantle
2 Intrusive igneous rock
3 Extrusive (volcanic) igneous rock
4 Sediment
5 Sedimentary rock
6 Metamorphic rock

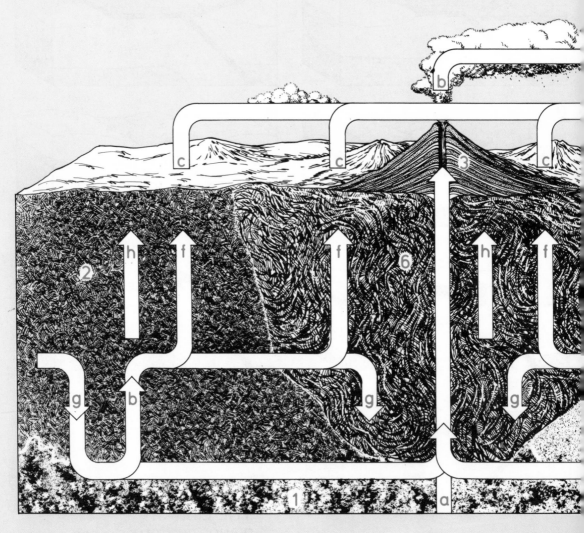

Colliding lithospheric plates thrust much of this above the sea. Subducting lithospheric plates bear igneous and sedimentary rock down into the mantle where heat and pressure turn it into metamorphic rock. Molten igneous rock rises through the cooler, denser rocks above, creating island arcs, injecting new material into the continental crust, and baking preexisting rocks.

The rock cycle and its associated cycle of erosion involve the processes and products covered in the next seven chapters.

Processes
a Emplacement
b Solidification
c Erosion
d Deposition
e Lithification (rock formation)
f Metamorphism
g Fusion
h Uplift

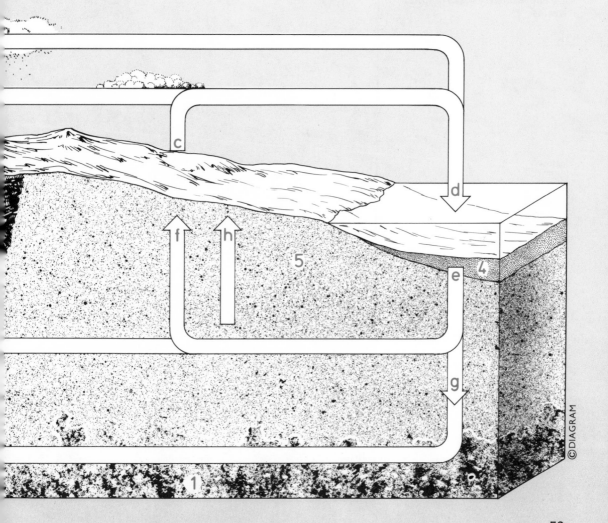

©DIAGRAM

Chapter 3 — FIERY ROCKS

Molten rock welling up from deep down in the Earth's interior cools and hardens at or near its surface to create such rocks as lavas and granite. Igneous or "fiery" rocks like these hold minerals that form the raw materials from which all crustal rocks derive. These pages describe major types and forms of igneous rocks, their ingredients, and the phenomena they yield – from mighty sheets and domes to tall volcanic cones, hot springs, and geysers. This chapter ends with a glimpse of fiery rocks of other worlds.

An example of columnar basalt showing the characteristic hexagonal form. (19th century engraving from *The National Encyclopaedia*)

Rocks from magma

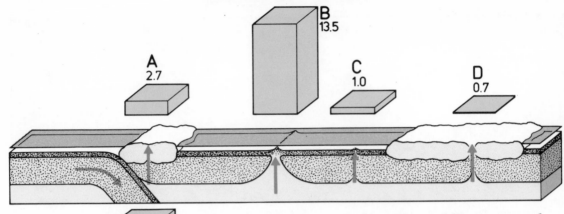

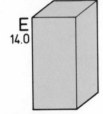

Magma production (above)
Columns show annual magma output and loss in cu km from specific areas.
A Destructive plate boundaries (subduction zones)
B Constructive plate boundaries (spreading ridges)
C Within oceanic plates
D Within continental plates
E Plate material consumed at destructive plate boundaries

Igneous or "fiery" rocks floor the world's oceans and form rock masses rising from the roots of continents. Such rocks arise directly from the molten underground rock material, magma. Magma occurs where heat melts parts of the Earth's upper mantle and lower crust. Most magma that has cooled and solidified escaped up through the crust from oceanic spreading ridges. Smaller quantities came from destructive plate boundaries and colliding continents.

Igneous rocks' ingredients (right)
Percentages of pale and dark minerals in coarse-grained and fine-grained rocks.

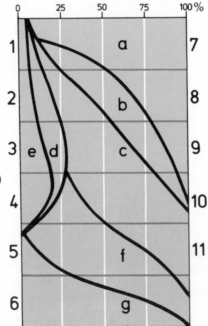

Coarse-grained rocks:
1 Syenite **2** Granite
3 Granodiorite **4** Diorite
5 Gabbro **6** Peridotite

Pale and dark minerals:
a Orthoclase feldspar
b Quartz
c Plagioclase feldspar
d Biotite **e** Amphibole
f Pyroxene **g** Olivine

Fine-grained rocks:
7 Trachyte **8** Rhyolite
9 Rhyodacite **10** Andesite
11 Basalt

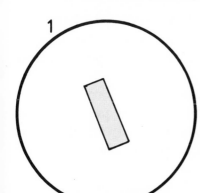

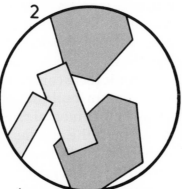

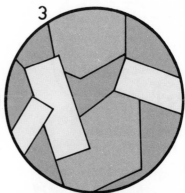

Igneous rocks hold many minerals. These are chiefly silicates (silicon and oxygen usually combined with a base or metal). The major silicates are feldspars: silicates of aluminum combined with certain other elements, notably potassium (in alkali feldspars) and sodium and/or calcium (in plagioclase feldspars). Other silicates include ferromagnesian minerals (rich in iron and magnesium), for instance amphibole, biotite mica, olivine, and pyroxene. Quartz is the sole silicate comprising only silicon and oxygen.

Silica-rich igneous rocks are described as "acid." In descending order of silica content the other major groups are intermediate, basic (or mafic), and ultrabasic (or ultramafic).

Which type of rock evolves depends upon the type of parent magma and the processes this underwent: absorbing magma-melted rock, losing gas, and cooling down. Different minerals "freeze out" or crystallize and separate at different temperatures, so some magma yields rocks with layers of different minerals. But an igneous rock usually contains one set of minerals that solidified at roughly the same temperature. Rate of cooling influences crystal size: slow cooling gives large mineral crystals; fast cooling yields fine-grained rocks.

The next pages describe igneous rocks formed in different conditions.

Crystals taking shape (above)
Minerals solidifying in sequence give igneous rocks an interlocking crystal texture.
1 First mineral starts forming in molten magma.
2 Second mineral forming.
3 Last mineral to form fills any remaining space.

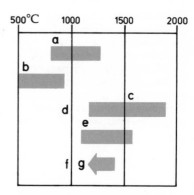

Solidifying minerals (above)
Temperatures in degrees Centigrade at which 7 minerals crystallize:
a Amphibole
b Quartz
c Olivine
d Mica
e Pyroxene
f Orthoclase feldspar
g Plagioclase feldspar

©DIAGRAM

57

Fiery rocks formed underground

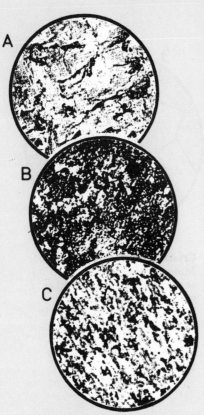

Intrusive igneous rocks are those produced where magma cools and hardens underground. Geologists place intrusive rocks in two categories: plutonic and hypabyssal.

Plutonic rocks comprise great masses formed deep in mountain-building zones, some by partial fusion of lower continental crust, and some from magma rising from the mantle. Slow cooling gives big mineral crystals, thus coarse-textured rocks including (acid) granite and granodiorite, (intermediate) syenite and diorite, (basic) gabbro, and (ultrabasic) peridotite. Granite – mainly made of quartz, feldspar, and mica – is the chief igneous rock of continental crust. Erosion of overlying rock exposes granite masses like the domes above California's Yosemite Valley and the tors of Dartmoor in southwest England.

Hypabyssal rocks are relatively smaller masses, often strips or sheets. Such rocks cooled at a lesser depth and faster than plutonic rock, so hold smaller

Three intrusive rocks (above)
A Granite, an acid rock rich in quartz and feldspar
B Diorite, an intermediate rock, mainly plagioclase feldspar with darker minerals such as biotite or hornblende
C Gabbro, a basic rock, mainly plagioclase feldspar and pyroxene

How granite forms (right)
Light, molten blobs of granite rise from rocks melted deep down in the crust. Great blobs called plutons coalesce and cool as batholiths – immense rock masses in the cores of mountain ranges like the Sierra Nevada of California.

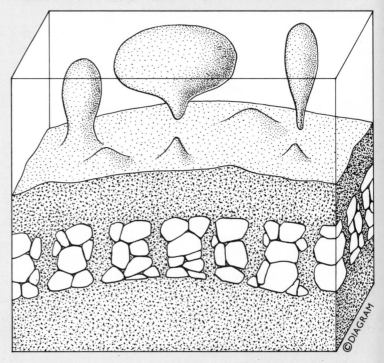

crystals. Hypabyssal rocks include (acid) microgranite and microgranodiorite, (intermediate) microsyenite and microdiorite, and (basic) diabase (dolerite).

Intrusive rocks produce these features:

1 Batholith A huge deep-seated, dome-shaped intrusion, usually of acid igneous rock.

2 Stock Like a batholith but smaller; (irregular) surface area under 40sq mi (about 100sq km).

3 Boss A small circular-surfaced igneous intrusion less than 16mi (26km) across.

4 Dike A wall of usually basic igneous rock such as diabase (dolerite) injected up through a vertical crack in preexisting rock.

5 Sill A sheet of usually basic igneous rock intruded horizontally between rock layers.

6 Laccolith A lens-shaped usually acidic igneous intrusion that domes overlying strata.

7 Lopolith A saucer-shaped intrusion between rock strata; up to hundreds of miles across.

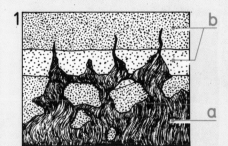

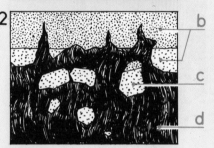

Stoping (above)
1, 2: Granitic magma (**a**) rises by melting crustal rock (**b**). Xenoliths (**c**) are lumps of crust embedded in the granite (**d**).

Intrusive rocks (left)
1 Batholith
2 Stock
3 Boss
4 Dike
5 Sill
6 Laccolith
7 Lopolith

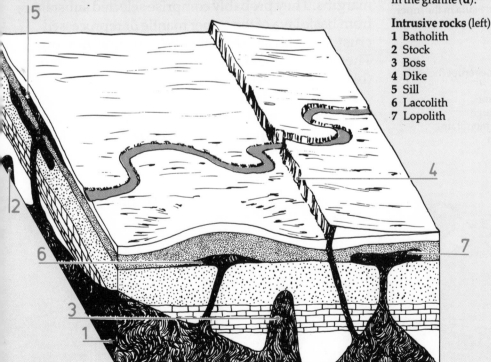

Volcanic rocks

Extrusive, or volcanic, igneous rocks occur chiefly at volcanic vents along the active margins of lithospheric plates. Here, magma erupts as lava, which cools and hardens quickly on the surface as fine-grained or glassy rock.

Basic lavas are rich in metallic elements but poor in silica. They flow easily and erupt relatively gently. The best-known product is basalt, which accounts for more than 90 per cent of all volcanic rock. This dark, fine-grained rock contains the minerals plagioclase feldspar, pyroxene, olivine and magnetite. Basalt is formed by partial melting of peridotite, the chief rock of the upper mantle. Basalt wells up from oceanic spreading ridges and builds new ocean floor. More appears in continental rift valleys, and rows of volcanoes like the Hawaiian Islands – products of a fixed plume of magma punching through the Pacific plate moving over it.

Acid (silica-rich) lavas appear at destructive plate margins. They probably comprise selected substances from basic lava of the upper mantle or reprocessed crust. Acid lavas are explosive and slow-flowing. They produce such rocks as dacite, rhyolite, and (black, glassy) obsidian.

Intermediate lavas contain plagioclase feldspar and amphibole, sometimes also alkali feldspar and quartz. They stem from partial melting of certain minerals in

Basalt columns (above)
As basalt lava cools it shrinks and sometimes splits into vertical columns. Famous examples include Ireland's Giant's Causeway, and Staffa in the Inner Hebrides.

Where volcanoes erupt (below)
a Island arc
b Shield volcano
c Spreading ridge
d Cordilleran mountains
e Lava plateau
f Rift valley

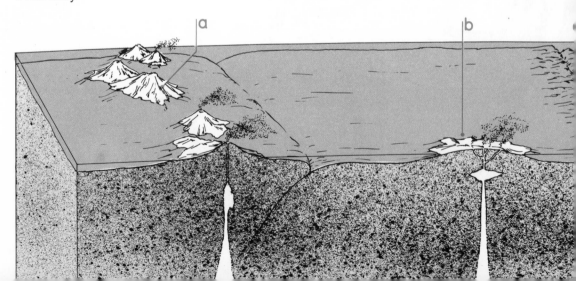

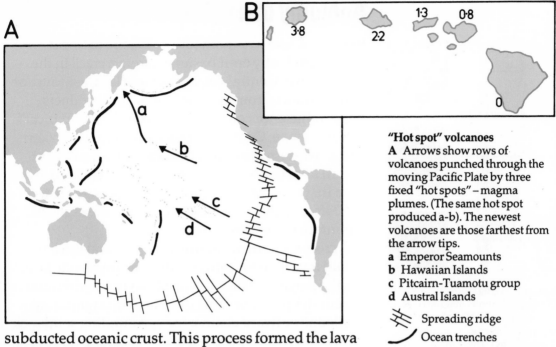

A

B

3·8   2·2   1·3   0·8   0

"Hot spot" volcanoes

A Arrows show rows of volcanoes punched through the moving Pacific Plate by three fixed "hot spots" – magma plumes. (The same hot spot produced a-b). The newest volcanoes are those farthest from the arrow tips.
a Emperor Seamounts
b Hawaiian Islands
c Pitcairn-Tuamotu group
d Austral Islands

〰️ Spreading ridge
⌒ Ocean trenches

B Hawaiian Islands. Numbers are ages of the youngest volcanic rocks in millions of years.

subducted oceanic crust. This process formed the lava andesite, named for the Andes Mountains. Found on the landward side of oceanic trenches, andesite builds island-arc volcanoes and tacks new land onto the rims of continents.

More than 850 known volcanoes have erupted in the last 2000 years. Those emitting continuously or from time to time are active; many form a "Ring of Fire" around the Pacific Ocean. Volcanoes not erupting in recent times are known as dormant. Long-inactive volcanoes are said to be extinct; some occur where colliding plates fused together many million years ago.

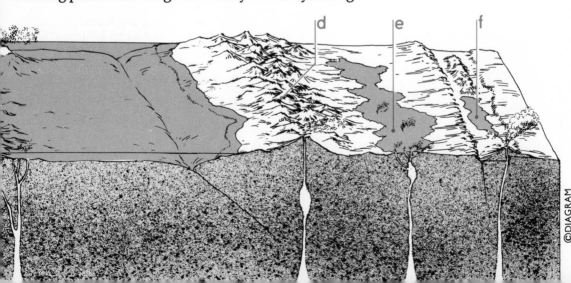

©DIAGRAM

Anatomy of a volcano

Volcanoes take two main forms. Fissure or linear volcanoes chiefly emit basic lava from a crack in the Earth's crust. Central volcanoes yield lava, ash, and/or other products from a single hole. These products build a shield- or cone-shaped mound – the "typical" volcano shape. Central volcanoes can grow high and fast. In western Mexico in 1943 Paricutín grew 490ft (150m) high in a week and reached 1500ft (450m) in a year. In western Argentina extinct Aconcagua towers 22,834ft (6960m) above sea level; this is the highest mountain in the western hemisphere.

A cross section through an active central volcano would reveal these features. Miles below the surface lies the magma chamber, a reservoir of gas-rich molten rock under pressure. This pressurized magma may "balloon" outward against the surrounding solid rock until it can relieve the pressure by escaping through a weakness in the crust above. From the chamber, magma then rises through a central conduit. As magma rises the pressure on it is reduced and its dissolved gases are freed as expanding bubbles. Finally the force of gases blasts open a circular vent on the Earth's surface. From this outlet ash, cinders and flows of lava build the main volcano shield or cone. Vent explosions shape its top as an inverted cone or crater. Meanwhile, side vents on the flanks of the volcano release ash or lava that may build subsidiary cones.

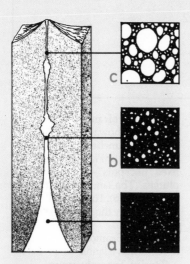

Gas and magma (above)
Inside a volcano as magma rises its pressure falls. Dissolved gases escape and form expanding bubbles (**abc**). These force magma out of the volcano.

How Paricutín grew (below)
This diagram shows the rapid growth of Paricutín, Mexico. (Heights are in meters, distances in kilometers from the center of the cone.)
a After 24 hours
b After one week
c After one year
d Paricutín village church

Volcanic features (right)
a Magma **b** Magma chamber
c Pipe **d** Side vents
e Vent **f** Cone
g Subsidiary cone **h** Gas
i Ash **j** Lava

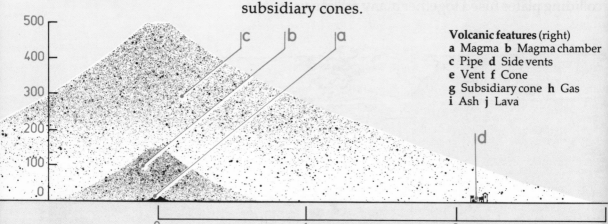

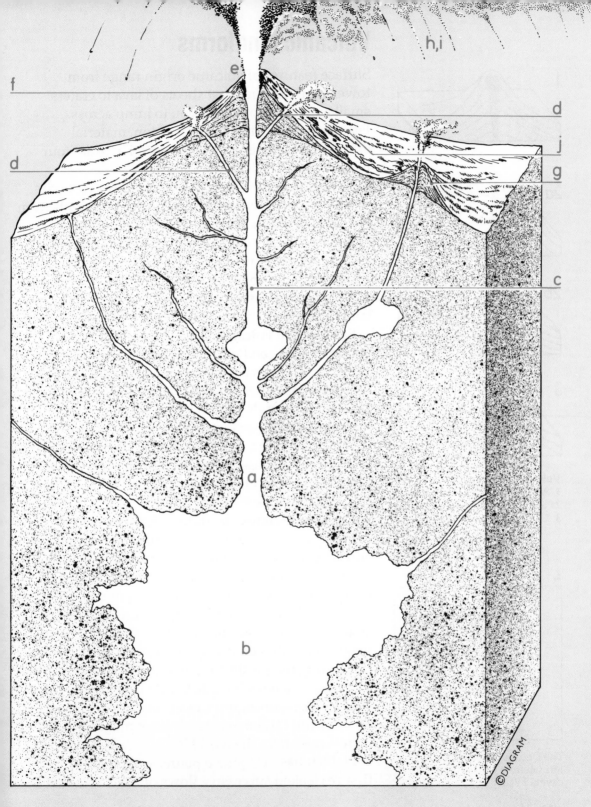

f

e

d

d

j

g

c

a

b

©DIAGRAM

Volcanic landforms

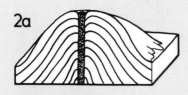

1

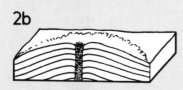

2a

2b

3

Volcano types (above)
1 Ash and cinder cone
2a and b Lava cones
3 Composite cone

The Deccan Plateau (above), formed from lava outflows, covers 250,000 square miles.

Surface features of volcanic origin range from towering peaks and vast sheets of lava to craters small and low enough for you to jump across. Features vary with type of eruption, material erupted, and the effects of erosion. There are four major types of volcanic landform:

1 **Ash and cinder cones**, or explosion cones, occur where explosive eruption ejects solid fragments from a central crater and/or subsidiary craters. The resulting concave cone is seldom higher than 1000ft (300m). Idaho's Craters of the Moon district has many such examples.

2 **Lava "cones"** usually form from slowly upwelling lava. They come in two main types:

a **Steep sided volcanoes** like France's Puy de Dôme and Lassen Peak in California grew from sticky acid lava that soon hardened. Squeezed out like toothpaste, very viscous lava built spines like Wyoming's Devils Tower.

b **Shield volcanoes** (gently sloping domes) formed from runny lava that flowed far before it hardened. From a seabed base 500mi (800km) across, Hawaii's Mauna Loa rises gradually 32,000ft (9750m) to the broad crater at its summit.

3 **Composite cones**, or strato-volcanoes, with concave, cone-shaped sides, feature alternating ash and lava layers. Composite cones account for most highest volcanoes. Fujiyama, Kilimanjaro, Mt. Rainier, and Vesuvius are four well-known examples. If solid lava plugs the main pipe to the crater, pent-up gases may blast the top off. If the magma chamber empties, the summit may collapse. Either way the product is a vast shallow cavity called a caldera. Calderas include Crater Lake in Oregon, Tanzania's Ngorongoro Crater, and Japan's Aso, whose 71mi (112km) circumference makes this the largest caldera in the world.

4 **Plateau basalts**, or lava plains, occur where fissures leaked successive flows of basic lava that

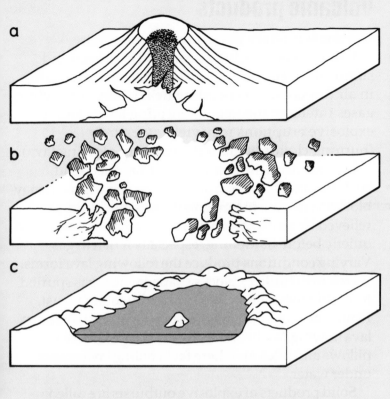

Volcanic basins (calderas)
These may be formed in several
ways, one of which is shown
(left).
a A lava plug bottles up
explosive gases below
b In time, pent-up gas pressure
blasts off the top of the volcano
c The explosion leaves a great
shallow cavity – a caldera or
basal wreck

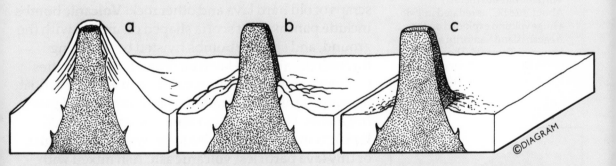

have blanketed huge areas in basalt. Basalt up to
7000ft (2100m) thick covers 250,000 sq mi (650,000 sq
km) in India's Deccan Plateau. Other outflows form
the US Columbia River Plateau, South America's
Paraná Plateau, the Abyssinian Plateau, and
Northern Ireland's Antrim Plateau.

Volcanic plug (above)
a Upwelling lava fills the
original volcano's central pipe
b Erosion attacks the soft outer
slopes
c Only the resistant lava plug
remains

©DIAGRAM

65

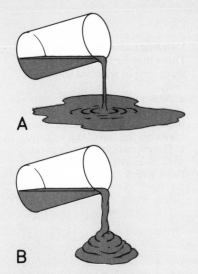

How lavas behave (above)
Lavas vary in viscosity –
resistance to internal flow.
A Low-viscosity lava flows
readily, like water.
B High-viscosity lava flows
sluggishly, like molasses.

Volcanic fall-out (below)
About 1628 BC, as redated in 1988,
a huge volcano exploded on the
Aegean island Santorini also
known as Thera (**a**). The ash-fall
(shown here tinted) covered
much of Crete (**b**) and might have
helped destroy Crete's great
Minoan culture.

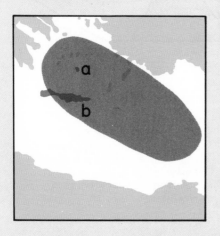

Volcanic products

Volcanoes produce gases, liquids, and solids.

Volcanic gases include steam and hydrogen and sulfur as well as carbon dioxide. Steam condensing in air forms clouds shedding heavy rain. Interacting gases intensify the heat in erupting lavas, and explosive eruptions may yield *nuées ardentes* (burning clouds of gas with scraps of glowing lava).

Liquid lava is the main volcanic product. Acid, sticky lava cools and hardens before flowing far. It may block a vent, causing magmatic pressure build-up, relieved by an explosion. Basic, fluid lava flows far and quietly before hardening, especially if rich in gas. Varying conditions produce the following lava forms. **Aa** features clinkery blocks shaped where gas spurted from sluggish molten rock capped by cooling crust. **Pahoehoe** has a skin dragged into wrinkles by molten lava flowing fast below it. **Pillow lava** resembles pillows and piles up where fast-cooling lava erupts under water.

Solid products of explosive outbursts are called **pyroclasts**. These can be fresh material or ejected scraps of old hard lava and other rock. **Volcanic bombs** include pancake-flat scoria shaped on impact with the ground, and spindle bombs twisted by whizzing through the air. Acid lava full of gas-formed cavities produces **pumice**, a volcanic rock light enough to float on water. **Ignimbrite** contains naturally welded glassy fragments. Hurled-out cinder fragments are called **lapilli**. Some volcanoes also spew vast clouds of **dust**, or tiny lava particles: **volcanic ash**. Ash mixed with heavy rain produces mudflows like the one that buried the Roman town Herculaneum when Mt. Vesuvius erupted in AD79. Immense explosions can smother land for miles around in ash and hurl vast quantities of dust into the higher atmosphere, cooling climates on a global scale and adding layers to deep-ocean sediments.

Violent eruptions destroy towns and farms. But volcanic ash provides rich soil for crops.

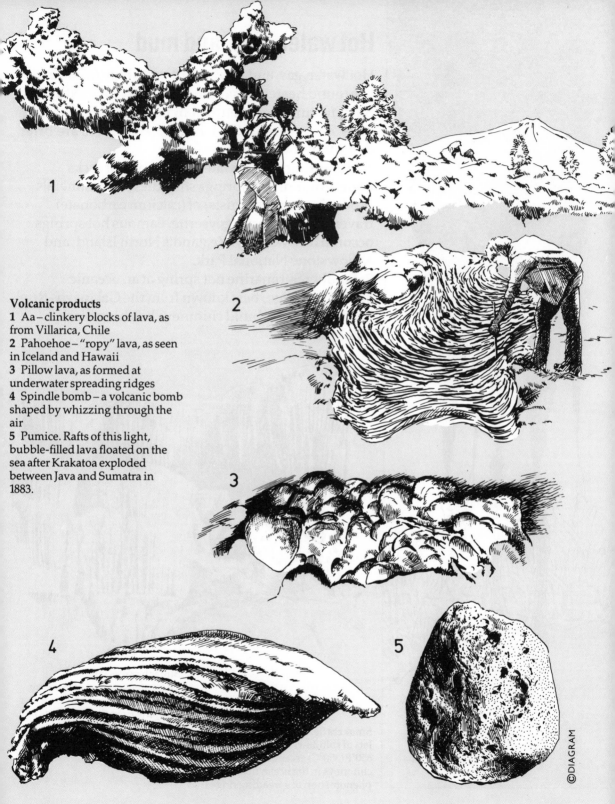

Volcanic products
1 Aa – clinkery blocks of lava, as from Villarica, Chile
2 Pahoehoe – "ropy" lava, as seen in Iceland and Hawaii
3 Pillow lava, as formed at underwater spreading ridges
4 Spindle bomb – a volcanic bomb shaped by whizzing through the air
5 Pumice. Rafts of this light, bubble-filled lava floated on the sea after Krakatoa exploded between Java and Sumatra in 1883.

©DIAGRAM

Hot water, gas, and mud

Hot water, gas, and mud squirt or dribble from vents in ground heated by volcanoes, mostly near extinction. Such features are plentiful in parts of Italy, Iceland, New Zealand, and the United States. Here are brief definitions of these forms.

1 Hot spring Spring water heated by hot rocks underground. Hot springs shed dissolved minerals producing sinters (crusts) of (calcium carbonate) travertine or (quartz) geyserite. Famous hot springs occur in Iceland, New Zealand's North Island, and Yellowstone National Park.

2 Smoker Submarine hot spring at an oceanic spreading ridge; best known from the Galapágos Rise. Emitted sulfides build chimneys belching black, smoky clouds.

Sinter terraces (above)
These steps consist of minerals deposited by mineral-rich water escaping from the ground as hot springs.

Smokers (left)
Jets of sulfide-rich water heated to 650°F (350°C) escape from mineral chimneys in the ocean floor – a phenomenon of spreading ridges.

3 Geyser Periodic fountain of steam and hot water forced up from a vent by water superheated in a pipe deep down. Famous geysers occur in Iceland and Yellowstone National Park.

4 Mud volcano Low mud cone deposited by mud-rich water escaping from a vent. Iceland, New Zealand's North Island, and Sicily have mud volcanoes.

5 Fumarole Small vent emitting jets of steam, as at Mt. Etna, Sicily, and in Alaska's Valley of Ten Thousand Smokes.

6 Solfatara Volcanic vent emitting steam and sulfurous gas; named after one near Naples, Italy.

7 Mofette Small vent emitting gases including carbon dioxide. Examples occur in France (Auvergne), Italy, and Java.

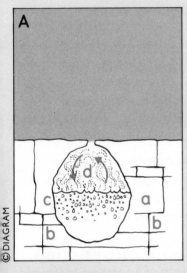

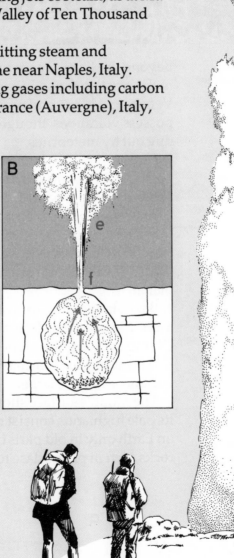

Geysers in action
Two diagrams show how one type of geyser operates.
A Hot rocks (**a**) heat water that flows through joints (**b**) and boils in a cave (**c**). Steam (**d**) collects under pressure.
B Pressure forces a jet of steam and water (**e**) from the cave's narrow mouth (**f**).

Fiery rocks of other worlds

The surface of Venus (above)
As seen from Russian space probe
Venera 14, the Venusian surface
revealed layered, weathered,
basaltic rocks similar to ones on
Earth.

Beyond the Earth, space probes reveal an igneous
rocky crust on the solid-surfaced planets Mercury,
Venus, and Mars, and on some moons. Certain worlds
possess volcanoes, though most of their craters were
dug out by meteorites.

Much-cratered Mercury includes smooth plains,
perhaps old basalt lava flows.

Basalt lava probably built the plains that largely
surface Venus. Here, radar shows vast twin cones
called Rhea Mons and Theia Mons. Perhaps the largest
(shield) volcanoes in the solar system, they may
contribute to the sulfuric-acid clouds that cloak this
planet.

Basalt lava plains sprawl over much of northern
Mars. Mars also has four mighty, very old, volcanoes.
Olympus Mons, the highest, towers at least 14mi
(23km) above the Martian plains. Only Venus's giant
volcanoes may be larger.

The Moon's surface shows two major types of rock.
Its pale highlands consist mainly of anorthosite (found
on Earth only in old parts of continents) and related
rocks, rich in plagioclase feldspar. The Moon's dark

Mars' giant mountain (below)
This diagram contrasts the
relative heights and sizes of three
volcanoes:
a Olympus Mons, Mars
b Everest (Earth's highest peak
above sea level)
c Mauna Loa, Hawaii. Most lies
below the Pacific Ocean

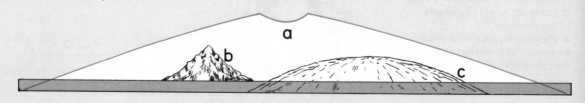

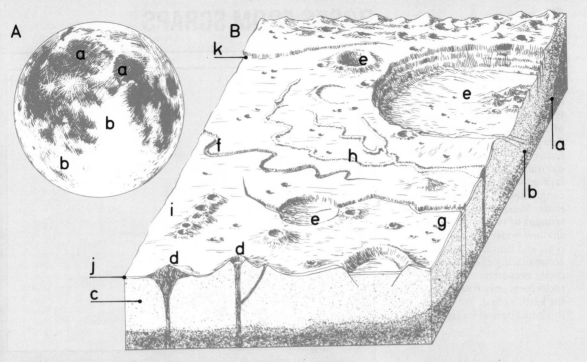

"seas" or maria are ancient basalt lava flows. These welled up, filling basins gouged out by the impact of asteroids or mini-moons. The Moon is now volcanically dead. But meteorite impacts melting surface rocks created breccias consisting of sharp stones in a glassy matrix.

The solar system's most volcanically active world is Io – a moon of Jupiter the same size as our Moon. Space probe images show hundreds of volcanic craters, and some immense volcanoes. Hot rocks heat sulfur in pockets underground. The molten sulfur rises, melts sulfur dioxide in pipes above, then squirts it out like water from a geyser. Plumes of sulfur compounds reach 200mi (300km) above the surface. Then they splash back, coloring the surface yellow, orange, black, and white like some colossal pizza.

Face of the Moon (above)
A The Moon as seen from Earth, reveals dark basalt plains (**a**), and pale highlands (**b**), largely of anorthosite.
B A block diagram reveals features of a basalt plain.
a Ancient crust (anorthosite)
b Breccia (rocks shattered by meteorite "bombs")
c Lava flows **d** Volcanoes
e Impact craters
f Sinuous rille (collapsed lava tunnel?)
g Linear rille (shallow rift valley)
h Wrinkle ridge **i** Crater chain
j Regolith (surface debris)
k Fault scarp

Sulfur fountains (right)
Molten sulfur compounds spurt high above the ever-active surface of a strange moon, Io.

Chapter 4 ROCKS FROM SCRAPS

When weather breaks up igneous or other surface rocks, wind and rivers bear away their broken scraps. Most settle on the seabeds fringing continents. Heavy pebbles get washed just offshore, sand farther out, light, fine particles of silt and clay farther still. Pressure and natural cements convert these layered sediments to the rocks conglomerate, sandstone, and shale. Meanwhile compacted remains of swamp plants and shallow-water animals form coals and limestones. Elsewhere accumulating chemical deposits create evaporites. Sedimentary rocks form only five per cent of the Earth's crust, but they cover three-quarters of its land.

1 A coquina quarry on Anastasia Island, Florida. Coquina, a white limestone formed of broken shells and corals, is often used as a building material.
2 The white cliffs along the southeast coast of England were formed from chalk, a soft limestone chiefly composed of the shells of coccoliths.
(Engraving of coquina quarry from *Picturesque America* 1894. Pictures of chalk cliffs from the Mansell Collection)

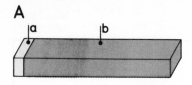

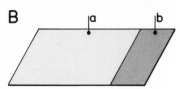

Rocks from sediments

Sediment or sedimentary rock covers most ocean floor and three-quarters of the land. On land this skin is usually a few miles thick; but layers up to 19mi (30km) thick collect in offshore basins. Most sedimentary rock comes from scraps of older (igneous or other) rocks eroded from the land, carried into lakes or seas by rivers, deposited, and then consolidated in a solid mass. When parent rock breaks up its minerals behave in different ways. Some of the silicates (the main mineral ingredients of igneous rocks) dissolve; others – quartz, for one – endure; and weathering creates new minerals – especially the clays that bulk large in most sedimentary rock. Besides the clastic sedimentary rocks (rocks made from fragments) others come from chemical precipitates or the remains of living things.

Processes converting sediment to rock are known as diagenesis. Two main processes occur. As sediments pile up their pressure squeezes water from the sediments below and packs their particles together. Then, too, some minerals laid down between grains cement a mass of sediment together.

Rock production (above)
A Relative volumes of (**a**) sedimentary and changed sedimentary rocks and (**b**) igneous and changed igneous rocks in the Earth's crust contrast with:
B The same rocks' exposed areas.

Where sediments accumulate
a Alluvial fans **b** Glaciers
c Rivers **d** Dunes **e** Lakes
f Lagoons **g** Estuaries
h Deltas **i** Tidal flats
j Continental shelf
k Continental slope
l Abyss **m** Reefs

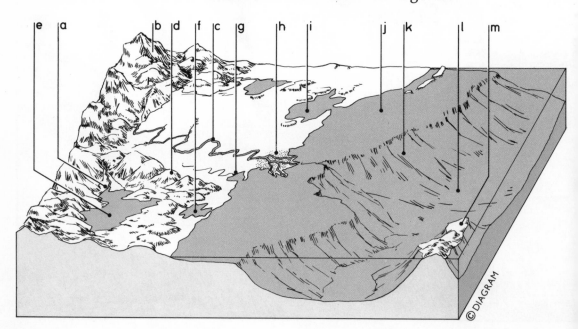

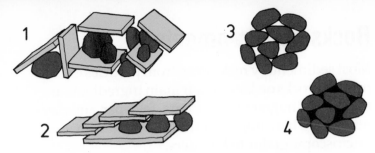

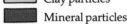

Changes converting sediment to rock leave traces in the finished product. Transportation of eroded sediments abrades and rounds their particles, sorts these by density or size, "rots" unstable minerals, and concentrates resistant minerals, including diamonds, and gold.

Deposition lays down sediments in broadly horizontal sheets called beds or strata, each separated from the next in the pile by a division called a bedding plane. Beds with ripple marks reveal ancient currents. Graded bedding (beds with grain size graded vertically) may hint at turbidity currents – sediment rich water sliding soupily down a continental slope. Cross bedding (sands laid down at an angle between two bedding planes) show features such as old dunes and sand-bars.

Two types of bedding (below)
A Graded bedding – where large particles had time to settle before small particles.
B Cross-bedding in sands laid down by migrating dunes or ripples.

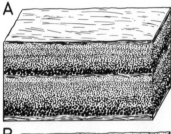

Cross-bedding in Utah (below)
A human figure shows the scale of cross-bedded sandstone sediments in Zion National Park.

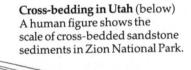

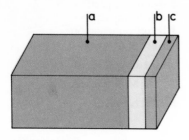

Sedimentary rocks (above)
Sedimentary rocks can be divided
into the following:
a Shale: 81 per cent
b Sandstone: 11 per cent
c Limestone: 8 per cent

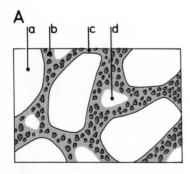

Clastic sedimentary rocks
A Above: Ingredients in a clastic
sedimentary rock.
a Clasts **b** Matrix
c Cement **d** Pore spaces
B Where rock-forming fragments
settle.
a River mouths (mud and sand)
b Inshore deposits (large sand
particles form sandstones)
c Offshore deposits (fine mud
particles form clay or shale)

Rocks from fragments 1

Most sedimentary rocks form from particles eroded
from the rocks on land. Their main ingredients are
clasts (rock fragments) of quartz, feldspar, and clay
minerals. These fragments range in size from
microscopic grains to boulders.

More than 90 per cent of all sedimentary rock
contains particles no bigger than a sand grain. Many
geologists classify such particles by size in two main
groups. The (fine-grained) lutites with grains less than
0.06mm diameter produce mudstone, siltstone, and
shale. The (medium-grained) arenites or sandstones
with grains of 0.06-2mm give arkose, graywacke, and
orthoquartzite. Here are brief descriptions of six fine-
and medium-grained rocks.

1 Mudstone Soft rock made of clay minerals of less
than 0.004mm diameter.

2 Siltstone Rock formed of particles 0.004-0.06mm
in diameter.

3 Shale Mudstone, siltstone, or similar fine-grained
rock of silt and clay split easily along its bedding
planes. Shale accounts for more than 80 per cent of all
sedimentary rock.

4 Orthoquartzite A "clean" or pure arenite mainly
made of quartz after other substances have been
removed. (Arenites account for more than 10 per
cent of all sedimentary rock.)

5 Arkose An arenite rich in feldspar derived from
gneiss or granite.

6 Graywacke A muddy, often grayish sandstone with
mixed-size particles including quartz, clay minerals,
and others.

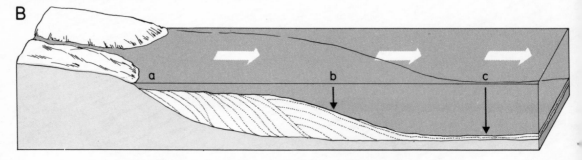

Four sandstones (right)
Variations in textural maturity
hint at places of origin.
A Immature (river floodplain)
B Submature (river or tidal
channel)
C Mature (beach)
D Supermature (desert dune)

Mineral grains

Clay matrix

Natural cement

Sandstones classified (right)
Varying proportions of four
components yield four types of
sandstone, or arenite.
Components: **a** Matrix
b Quartz **c** Rock fragments
d Feldspar
Sandstones: **1** Greywacke
2 Quartzite **3** Lithic arenite
4 Arkose

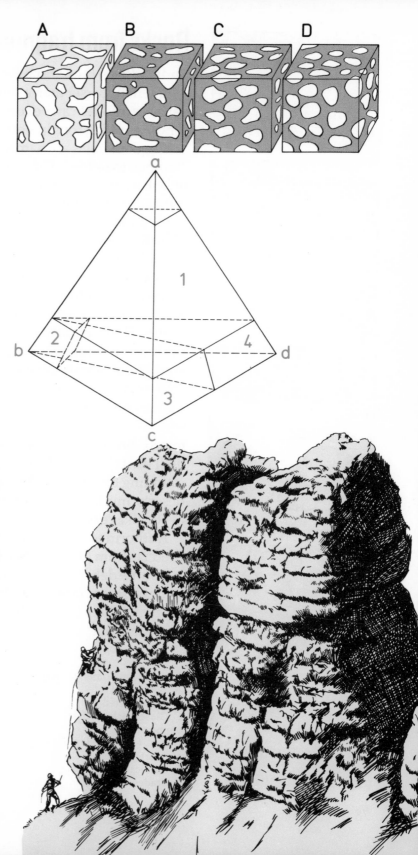

Shale (above)
An outcrop reveals bedding
planes in the fine-grained
sedimentary rock, shale.

Sandstone (right)
Weathering accentuates the
horizontal bedding planes and
vertical joints in these sandstone
masses.

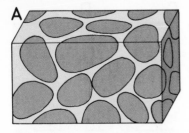

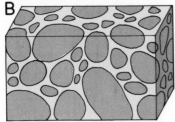

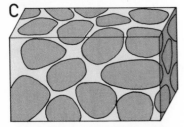

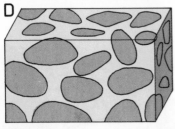

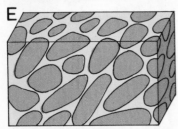

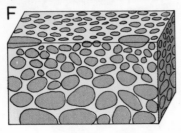

Rocks from fragments 2

Rudites (from the Latin *rudis*: "coarse") are clasts (rock fragments) coarser than a sand grain. Mixed with finer particles, rudites can be consolidated into natural concretes called conglomerates and breccias.

Conglomerates are named from the Latin for "lumped together." They contain rounded fragments – pebbles, cobbles, and/or boulders – and often represent waterborne and watersorted remnants of eroded mountain ranges or retreating rocky coasts. They accumulate along mountain fronts, in shallow coastal waters, and elsewhere; becoming mixed with sand, then bound by natural cement. How clasts in a conglomerate lie sorted, packed, and graded offers clues to how or where it was laid down. The thickest masses of conglomerate – as in the Siwalik Formation of the Himalayas' foothills – mark the aftermath of an orogeny.

Breccias (from the Italian for "rubble") are rocks containing sharp-edged, unworn, usually poorly sorted fragments, often embedded in a clay-rich matrix. Breccias form usually near their place of origin; their clasts have not been carried far enough to suffer rounding by abrasion. Many breccias originate in talus, deserts, mudslides, faulting, meteorite impact, or shrinkage of evaporite beds.

Authorities tend to separate conglomerates and breccias from tillites – poorly sorted, ice-eroded, ice-borne debris consolidated into solid rock. Many tillite clasts are faceted, with slightly rounded edges. Ancient tillites occur in South America, Africa, India, and Australia.

Types of conglomerate (left)
A Well sorted
B Poorly sorted
C Close-packed (clast supported)
D Loosely packed (matrix supported)
E Imbricated
F Graded bedding

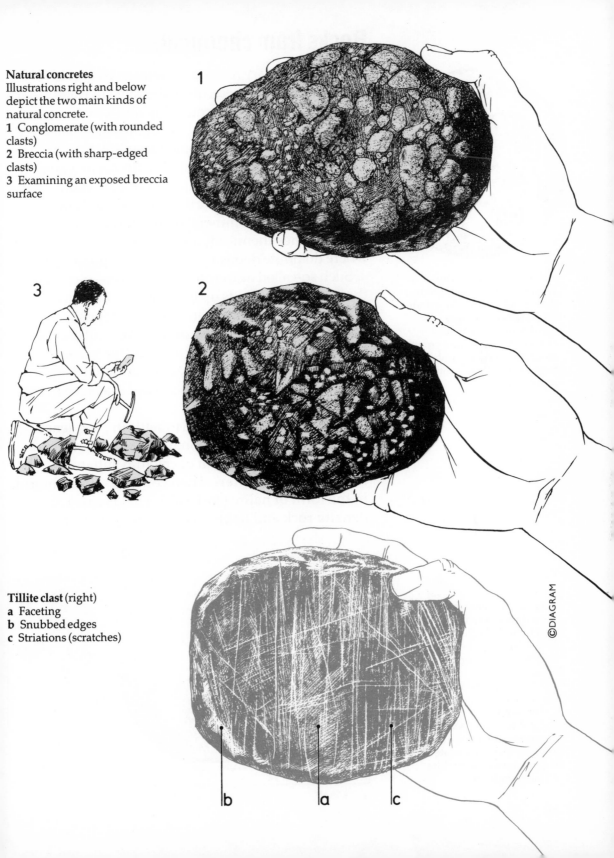

Natural concretes
Illustrations right and below depict the two main kinds of natural concrete.
1 Conglomerate (with rounded clasts)
2 Breccia (with sharp-edged clasts)
3 Examining an exposed breccia surface

Tillite clast (right)
a Faceting
b Snubbed edges
c Striations (scratches)

©DIAGRAM

Rocks from chemicals

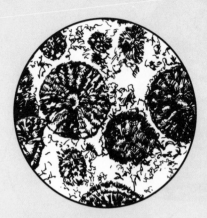

Oolites enlarged (above)
This much-magnified section through a Cambrian limestone shows spherical oolites (also called ooids) cemented by carbonate. Each oolite's growth around a sand grain or shell fragment produced a concentric, radial structure.

Some sedimentary rocks and minerals consist of chemicals that had been once dissolved in water.

Certain limestones formed this way. Oolitic limestone consists of billions of oolites: tiny balls produced by calcium carbonate accumulating on particles rolled around by gentle currents in warm, shallow seas. Oolite forms like this today on the Bahama Banks. Dolomitic limestone (limestone mainly made of the mineral dolomite) occurs where certain brines chemically alter preexisting limestone or where dolomite deposits form in an evaporating sea.

Such so-called evaporites underlie one-quarter of the continents in beds up to 4000ft (1220m) thick. Evaporites form now where chemical deposits accumulate in evaporating desert lakes and coastal salt flats. But certain old evaporites could have been precipitated from chemically oversaturated deep offshore waters of almost landlocked seas such as the Mediterranean.

Three main minerals tend to settle in a sequence. First comes calcium carbonate. Next is gypsum (a granular crystalline form of calcium sulfate combined with water). Then comes sodium chloride in the form of halite (rock salt). This is a soft, low density rock and liable to flow. Pressure from

Deep-water evaporites (right)
Some thick evaporite layers might have formed like this.
A In a deepwater sea basin evaporation outstrips freshwater input.
B A barrier sill partly shuts out the open sea.
C The basin water grows denser and sinks until evaporites are precipitated in this sequence:
a Dolomitized carbonate
b Gypsum with anhydrite
c Halite (rock salt)

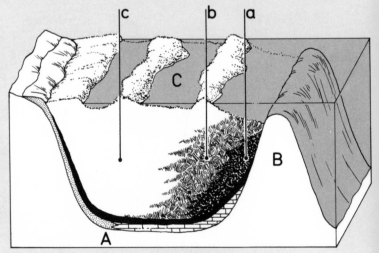

©DIAGRAM

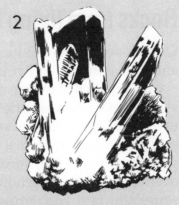

Two evaporites (left)
1 Dolomite, derived from small shells, and named after the French geologist Dieudonné Dolomieu (1750–1801)
2 Gypsum, a sulfate mineral named from *gypsos*, the Greek word for chalk.

overlying rocks forces up huge plugs or domes of salt beneath the coast of Texas and Louisiana, and in parts of Germany, Iran and Russia.

Besides the rocks and minerals just named, there are other chemical deposits. A few have or had important economic uses – particularly borax, chert and flint, certain iron-rich compounds, nitrates, and phosphorites. But some of these are partly biological in origin; and scientists disagree about how certain forms of iron occurred.

Salt dome (above)
a Salt core up to 6mi (10km) high
b Limestone-anhydrite cap rock
c Strata uplifted by salt core

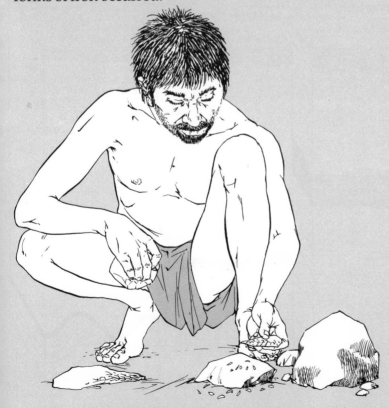

Flint knapping (left)
Stone Age man learnt that controlled blows could split this fine-grained quartz into sharp-edged weapons and tools.

Rocks from living things

Organic sediments produce the rocks we know as coals
and limestones. (For Petroleum see pp. 228-229.)
   Coals are rich in carbon derived from swampy
vegetation. Coal type varies with the processes
involved. Coal formation starts when plants die in wet
acid conditions; instead of rotting completely the
plants turn to the soft, fibrous substance peat. Later,
overlying sediments drive out much moisture and
squash the peat, converting it to lignite (soft brown
coal). Even greater pressure gives bituminous coal –
harder, blacker, and with a higher carbon content. The
final stage is anthracite – a hard, black, shiny coal with
the highest carbon content in the series. Most of the
world's coalmines tap remains of low-lying forests
drowned from time to time by an invading sea and
buried under sediments.

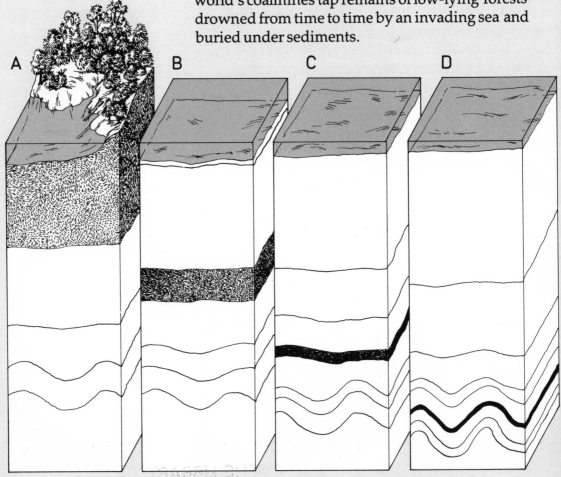

A   B   C   D

82

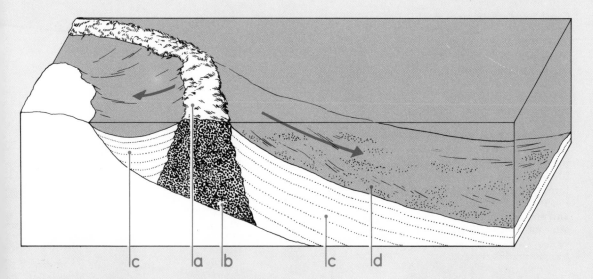

c a b c d

Limestones are rich in calcium and magnesium carbonates. They make up about 8 per cent of all sedimentary rock; only shale and sandstone are more plentiful. Organic limestones contain calcium carbonate extracted from seawater by plants and animals that used this compound for protective shells. These rocks include reef limestones built up from the stony skeletons of billions of coral polyps and algae inhabiting the beds of shallow seas. Coquina is a cemented mass of shelly debris. Chalk is a white, powdery, porous limestone comprising tiny shells of fossil microorganisms, drifting in the surface waters before they died and rained down on the bottom of the sea.

Origins of limestone (above)
Reef limestone forms in shallow waters from these four ingredients.
a Living corals of the upper reef, just below the sea
b Dead coral base of reef
c Coral broken from the reef
d Debris from the reef, remains of deeper-water animals and plants

Three organic rocks (below)
1 Bituminous coal (mainly carbon)
2 Coquina (shelly limestone)
3 Chalk (a powdery limestone)

1

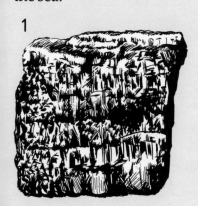

2

3

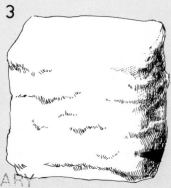

©DIAGRAM

Chapter 5

Great loads of sediment or ice depress tracts of continental crust. Where a load has been removed the land bobs up again. As lithospheric plates collide or split, tension or compression tilts, folds, squeezes, and breaks, the rigid rocks, producing folds, faults, and earthquakes. Then, too, meteorites hurtling from space punch craters in the crust, while blobs of molten rock push up through the crust from below. Events like these produce heat and pressure that deform and alter solid rocks. This chapter ends with an account of rocks transformed that way.

Scenes of destruction recorded after an earthquake hit San Francisco and Santa Rosa in 1906. (The Mansell Collection)

Rising and sinking rocks

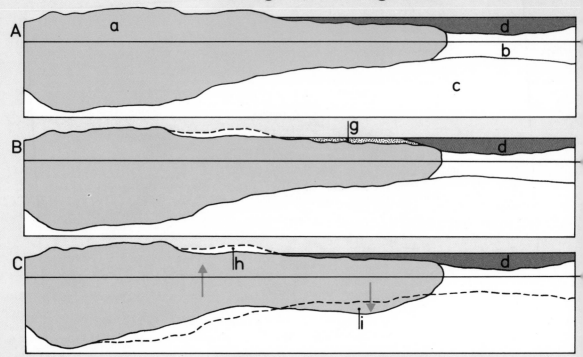

Maintaining balance (above)
A Crust in equilibrium
a Continental crust
b Oceanic crust
c Mantle **d** Sea
e Line of crustal equilibrium
B Disequilibrium due to land erosion (**f**), and deposition on continental shelf (**g**).
C Upwarping (**h**) and downwarping (**i**) restore equilibrium.

Great tracts of the Earth's crust slowly rise or sink. Block mountains and the Black Sea and Mediterranean Sea are products of such movements, called epeirogenesis. Its cause is some disturbance that affects isostasy – the state of balance of the Earth's crust floating on the denser underlying mantle. Crust is four-fifths as dense as mantle, so a mountain mass a mile high is "balanced" by four miles of crust below sea level.

Warping – the gentle rise or fall of crust – results from two pairs of processes: erosion and deposition, and freezing and melting.

Erosion that wears away a mountain mass reduces the weight pressing on the underlying crust, so the surface of the land bobs up. Such upwarping probably means that few large land surfaces get worn down level with the sea. Indeed vast slabs of Africa have been upraised as plateaus.

Deposition of eroded sediment depresses or downwarps deltas and large tracts of continental rim.

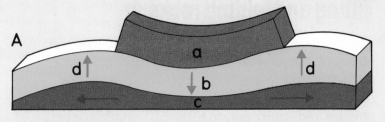

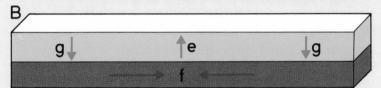

Sediments thousands of meters thick accumulate in
eugeosynclines – depressions in a continent's deep-
water margin. Shallow-water sediments including
limestones form long lenses in continental-shelf
depressions called miogeosynclines.

Growing ice sheets depress the crust beneath,
squeezing out asthenospheric material which pushes
up the crust beyond the ice sheet's rim. When ice
sheets melt, the land below bobs up again – a process
that has lasted 10,000 years or more in Scandinavia and
northeast Canada.

While isostatic change describes land moving up or
down, eustatic change is a worldwide shift in the level
of the sea. Eustatic change occurs if ocean basins grow
or shrink, or if they gain or lose sea water. Sea level falls
when much of the world's surface moisture becomes
locked up in ice sheets on the land. But water freed by
melting ice sheets lifts the level of the sea. Clues to
eustatic and isostatic change include the drowned
valleys and raised beaches of some coasts described in
Chapter Eight.

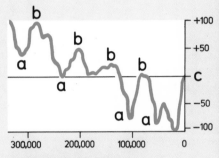

Changing sea level (above)
A graph depicts in meters likely
changes in level of the
Mediterranean in the last 300,000
years.
a Glacial periods
b Interglacials
c Present sea level

Downwarped crust (below)
A section New York (**A**)-Maine
(**B**) shows rocks up to 20,000ft
(6000m) thick produced where
offshore sediments downwarped
crust 500 million years ago.
1 Old miogeosyncline (shallow-
water limestones)
2 Old eugeosyncline (deeper-
water shales, etc)

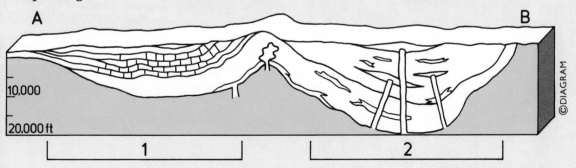

Tilting and folding rocks

Sediments and plateau basalts are laid down as horizontal beds or sheets. Old deposits lie almost undisturbed across great tracts of ancient, stable continental shields. But in unstable regions crustal tension and compression tilt, fold, squeeze, or break the level layers, and thrust them up as mountains. Which type of deformation happens depends on pressure, temperature, strain-rate (compression in a given time), and composition of affected rocks.

Folding occurs largely deep down along the edges of colliding continental plates. Here steady stress, and high temperatures and pressures make normally brittle rocks bend instead of breaking. Thus in quartzite, quartz grains slide about and dissolve at stress points. Such processes produced repeated folding in the Alps and Himalayas.

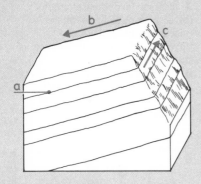

Tilted rocks (above)
Labels indicate key features of tilted rock beds.
a Bedding plane
b Dip
c Strike

Folded rocks exposed (above)
This Cornish sea cliff reveals rock layers doubled over in a small recumbent fold. Rock folds of almost any kind can span just a few feet as here, or measure miles across, as in some regions of the Alps.

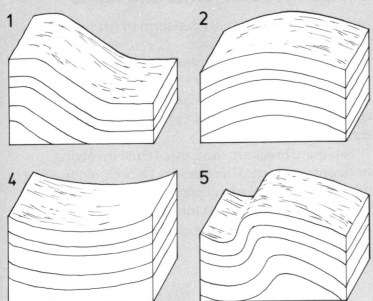

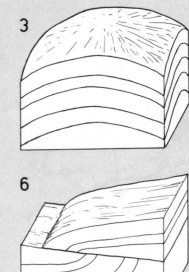

Folds take various forms, especially these.

1 Monocline A steep step-like fold, bounded by upper and lower bends in a set of rock layers.

2 Anticline Rock beds upfolded into an arch from a few feet to many miles across. Anticlines form much of the Jura Mountains of France and Switzerland. An anticline containing many lesser folds is an anticlinorium. Huge anticlines are geanticlines.

3 Pericline An anticline in the form of an elongated dome. The Bighorn Mountains of Wyoming show a periclinal structure.

4 Syncline Downfolded sedimentary rock layers that form a basin such as the London Basin. A syncline with subsidiary synclines is a synclinorium. Immense synclines are called geosynclines.

5 Overfold A lopsided anticline with one limb (side) forced over the other. Extreme overfolds are called recumbent folds.

6 Nappe A recumbent fold sheared through so that the upper limb is forced forward, perhaps many miles. Nappes feature prominently in the Alps. Nappes with rocks forced over each other in slices like a pack of cards are imbricate structures.

Six folds (above)
1 Monocline
2 Anticline
3 Pericline
4 Syncline
5 Overfold
6 Nappe

© DIAGRAM

89

Breaking rocks: joints and faults

Joints and faults are splits that form in stressed rock – particularly near the surface.

Joints are cracks with little movement of the rock on either side. They open up as cooling igneous rock contracts, and in other rock subject to tension or compression. Joints occur in parallel sets, sometimes at right angles to each other.

Faults are breaks in the Earth's crust involving horizontal or vertical movement, or both, along a line of weakness called a fault plane. Block-faulting – break-up of a slab of crust into fault-bounded blocks – creates some of the world's great valleys and upland areas.

Joints (above)
In many sedimentary rocks joints occur at right angles to bedding planes.
a Joints **b** Bedding planes

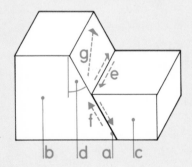

Anatomy of a fault (left)
a Fault
b Upthrow
c Downthrow
d Hade (inclination to vertical)
e Heave (lateral shift)
f Throw (vertical shift)
g Net movement

Rift Valley (below)
Such troughs lie between parallel faults with throws in opposite directions.

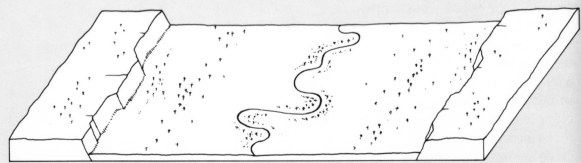

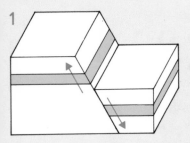

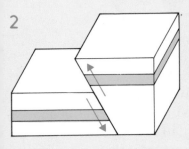

Here are six major types of fault or fault block.

1 Normal fault Stretching breaks rocks along a steep fault plane, and one block drops or rises against the other.

2 Reverse fault Compression forces one block up and over another. A thrust fault is a reverse fault with a low-angled fault plane producing great horizontal movement.

3 Tear fault (alias strike-slip, transcurrent, or wrench fault.) Horizontal shearing along a vertical fault plane, as in California's San Andreas Fault (see also pp. 92-93). Transform faults are tear faults at right angles to oceanic ridges.

4 Graben A long, narrow block sunk between two parallel faults. Such blocks form the upper Rhine Valley, East African rift valleys, oceanic spreading ridges' central rifts, and other rift valleys.

5 Horst A horizontal block raised between two normal faults. Examples are the Black Forest, Vosges, Korea, and Sinai.

6 Tilt block An uplifted, tilted block. Tilt-blocks form the US Basin-and-Range country, Arabian and Brazilian plateaus, and the Deccan of India.

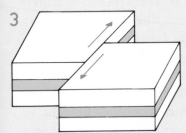

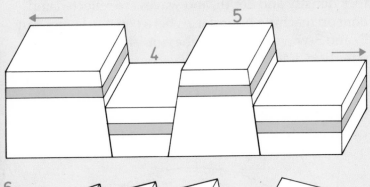

Types of fault (above/left)
1 Normal fault
2 Reverse fault
3 Tear fault
4 Graben
5 Horst
6 Tilt blocks

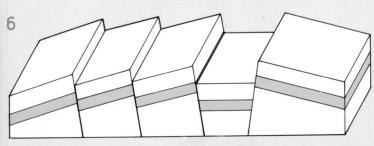

© DIAGRAM

Earthquakes

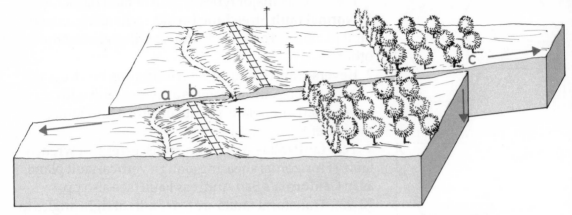

Evidence of earthquake (above)
Past 'quakes may leave such clues as these.
a River bend
b Displaced railroad
c Disrupted orchard

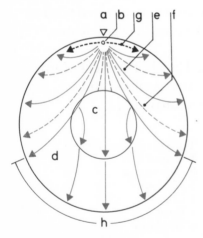

Seismic waves (above)
Waves reveal Earth's core.
a Epicenter (above focus)
b Focus **c** Core (blocks S-waves and deflects P-waves)
d Mantle **e** P-waves **f** S-waves
g L-waves **h** Shadow zones

Earthquake shocks (right)
Isoseismal lines link places with equal intensity of shock.
a Focus **b** Epicenter
c Isoseismal lines

An earthquake is a sudden shaking of the ground where stress-deformed rocks broke along a fault and now snap back into shape but in a new position.

An earthquake's point of fracture is its focus, which may be shallow, intermediate or deep – down to about 430mi (700km). From the focus, two types of seismic wave pass through the rocks. Compressional waves (alias primary or P-waves) produce push-pull forces. Distortional waves (alias secondary, shear, or S-waves) are slower and make rock particles oscillate at right angles to wave direction. Wave velocity increases with rock density and depth, and waves are reflected and bent on reaching boundaries between two layers. But P- and S-waves differ in behavior. This helped scientists detect the Conrad discontinuity between upper "granitic" and lower, denser, "gabbroic" crust; the Mohorovičić discontinuity between crust and upper mantle; and the Gutenberg discontinuity between the mantle and outer core (where S-waves

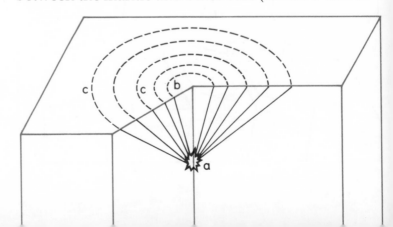

cease because they cannot pass through the liquid outer core).

At the surface are long (L) waves subdivided into Love waves vibrating horizontally at right angles to their direction, and Rayleigh waves that move through ground as waves move through the sea. By timing the arrival of these waves, three seismic stations can plot the epicenter (surface point above the focus).

L-waves spark off landslides, avalanches and other earthquake damage. Undersea earthquakes set off tsunami – mighty waves that devastate low coasts.

An earthquake's felt intensity is measured on the modified Mercalli scale, where 1 means "felt by few" but 12 means "damage total." The Richter scale measures magnitude, or energy released. Here each number stands for 30 times the energy of the number below; you would scarcely notice 2, but 8 would flatten a city. Many of the world's million earthquakes a year are fortunately slight.

Most occur at spreading ridges, oceanic trenches, and mountain-building zones. Strike-slip motion triggers earthquakes along California's notorious San Andreas Fault – a transform fault where western California moves northwest against the rest.

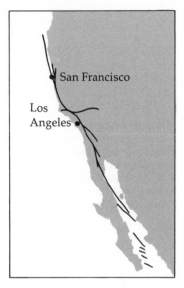

San Andreas Fault (above)
Lines show the San Andreas and associated faults. Land west of these is edging north west.

Earthquake belts (below)
These often coincide with active boundaries between lithospheric plates (see pp. 34-35.) Shallow earthquake foci lie 0-62mi (100km) down. Deep foci lie 62-435mi (100-700km) down.

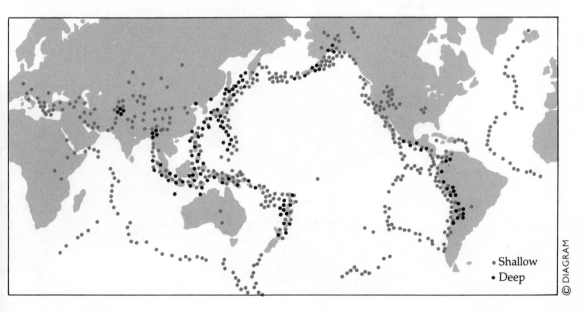

• Shallow
• Deep

© DIAGRAM

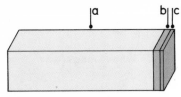

Meteorite percentages (above)
A bar diagram shows relative
abundance of the three main
types of meteorite.
a Stony: 93 per cent
b Iron: 5 per cent
c Stony iron: 2 per cent

Two meteorite types (below)
A Stony meteorite. Its white bits,
perhaps from an exploding star,
may be among the oldest solid
matter in our solar system.
B Iron meteorite, etched and
polished. Bands form a triangular
pattern of nickel-iron and other
nickel alloys.

Bombs from space

Whizzing specks, stones, and rocks bombard the
surface of the Earth from space. About 10,000 short tons
(9000 metric tons) shower down each year. These
missiles are meteorites – lumps weighing from a few
ounces up to 100 tons or more. A few are bits of planets
or the Moon struck off by other meteorites. Most are
fragments of colliding asteroids – a belt of so-called
minor planets between the orbits of the planets Mars
and Jupiter. Some meteorites are 4570 million years
old, and their ingredients hold clues to the solar
system's origin.

There are three main kinds of meteorite. Most are
silicate-rich **stony meteorites**. The bulk of these are
ordinary chondrites, containing solid silicate minerals.
Carbonaceous chondrites include organic compounds,
perhaps building blocks that made life possible.
Achondrites include basalt-like chunks, perhaps
volcanic rock from big asteroids. **Iron meteorites**, the
second major group, are iron with nickel. **Stony irons**,
the third group, contain roughly equal amounts of
nickel-iron and silicates.

Big stony meteorites break up and scatter before
they hit the ground. Big iron meteorites vaporize on
punching impact craters in the ground. Circular
depressions, shock structures in affected rocks, and

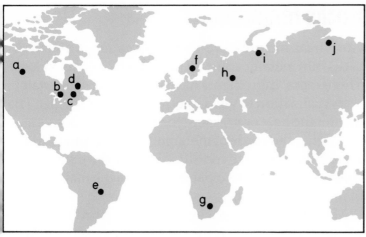

tell-tale iridium and nickel deposits have helped
scientists identify 90 sizable craters evidently gouged
by meteorites. They include Canada's vast
Manicouagan Crater 43mi (70km) across, Arizona's
famous Meteor Crater ¾mi (1.2km) across and 490ft
(150m) deep; and West Australia's Wolf Creek Crater
with a raised rim 100ft (30m) high.

More evidence for ancient impact lies in small, black,
glassy buttons, spheres, and teardrops collectively
called **tektites**. Found mostly in the Southern
Hemisphere, tektites were blobs of molten
sedimentary rock splashed high above the
atmosphere, before cooling, hardening and falling –
widely scattered – back to Earth.

Tektite (above)
Such glassy "buttons" come from
rocks melted and hurled high by
meteorites punching craters like
the one below.

Rocks remade 1

Great heat and pressure alter igneous and sedimentary rocks into metamorphic ("changed shape") rocks. These rocks' ingredients have undergone solid-state recrystallization to yield new textures or minerals. The greater the heat or pressure the greater the change and the higher the grade of metamorphic rock produced. Here we summarize conditions that induce such changes. The next two pages describe the major types of metamorphic rock.

Metamorphism takes several forms. Burial metamorphism affects the base of immensely thick layers of sedimentary rock – the deeper the burial, the greater the pressure. Dynamic metamorphism transforms rocks crushed against each other in fault zones. Retrogressive (from high to low-grade) metamorphism occurs at shear zones, for instance sub-oceanic transform faults where invading fluids introduce new elements that change rocks' chemical composition – a process known as metasomatism. Impact metamorphism affects rocks struck by meteorites. But the best-known agents of change are contact and regional metamorphism.

Heat and pressure (right)
This diagram relates rock-forming processes to temperature and pressure (indirectly shown by depth). Metamorphic rocks form in conditions between those producing sedimentary and igneous rocks.
a Diagenesis (sedimentary rock)
b Contact (thermal) metamorphism
c Burial metamorphism
d Regional metamorphism
e Anatexis or melting (igneous rock)

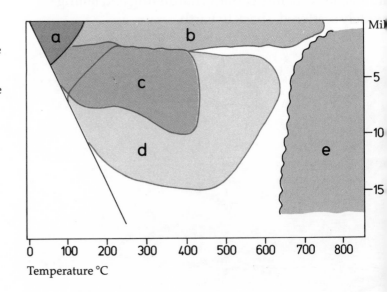

Temperature °C

Contact or thermal metamorphism occurs where a mass of magma invades and bakes country rocks (surrounding older rocks). Beyond a narrow baked zone extends a so-called contact aureole of altered rock. An injected granite batholith may change the rocks for several miles around.

Regional metamorphism covers much larger areas, downwarped where colliding continental plates built mountains. Intensely altered rocks produced this way show up in the exposed roots of old, eroded mountain ranges, as in the Canadian Shield, and parts of Scotland and Sweden; in the newer Alps and Himalayas; and in old subduction zones detectable in California, New Caledonia, and Papua New Guinea.

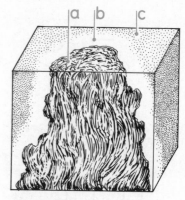

Contact aureole (above)
a Injected igneous mass
b Contact or metamorphic aureole of altered country rock.
c Unaltered country rock

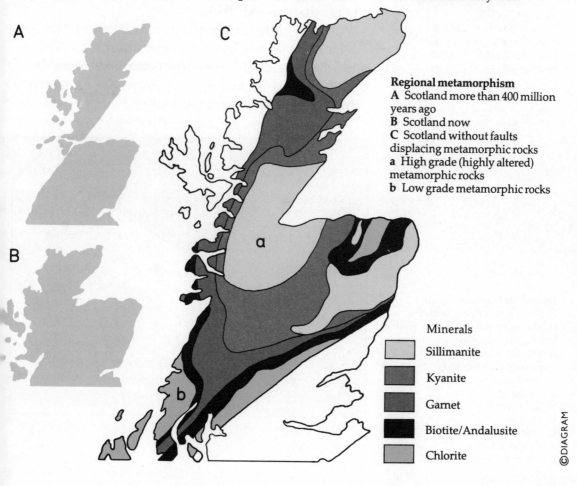

Regional metamorphism
A Scotland more than 400 million years ago
B Scotland now
C Scotland without faults displacing metamorphic rocks
a High grade (highly altered) metamorphic rocks
b Low grade metamorphic rocks

Minerals
☐ Sillimanite
☐ Kyanite
☐ Garnet
☐ Biotite/Andalusite
☐ Chlorite

©DIAGRAM

97

Rocks remade 2

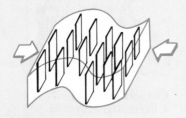

One type of sedimentary or igneous rock can produce a range of metamorphic rocks. Which occur depends on such variables as parent rock, amounts of heat, pressure, and fluids passing through the rock. Contact (thermal) metamorphism tends to produce fine-grained textures. Heat plus pressure, as in regional metamorphism, favor coarse-grained rocks with foliated minerals – minerals flattened and aligned in parallel bands at right angles to the stress applied.

Index minerals, ones formed at different temperatures or pressures, indicate a rock's metamorphic grade. Most metamorphic rocks are harder than sedimentary rocks, and pelitic (clay-rich) rocks undergo more change than basaltic rocks.

We show eight common types of metamorphic rock – 1-4 formed largely by contact metamorphism; 2 also and 5-8 by regional metamorphism. (Some types can be subdivided according to key minerals – for instance schist into talc schist and mica schist.)

Foliation (above)
1 Mica flakes haphazardly arranged in shale
2 The same flakes foliated – aligned by directed pressure – in slate, which splits along its foliation planes.

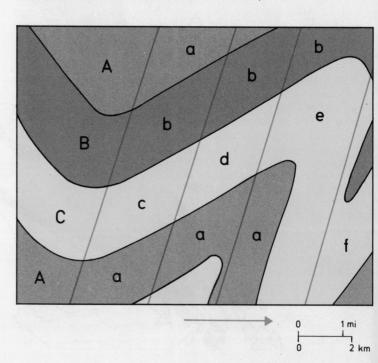

From shale to gneiss (right)
Zones of intensifying heat and pressure (the parallel bands, read left-right) change surface belts of sedimentary rocks (**A-C**) to metamorphic rocks (**a-f**).
A Sandstone
a Quartzite
B Limestone
b Marble
C Shale
c Slate **d** Phyllite
e Schist **f** Gneiss

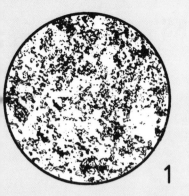

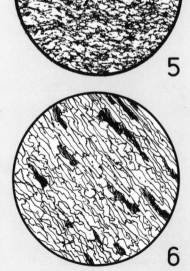

1 Hornfels Fine-grained, dark, flinty rock with randomly arranged minerals; formed from mudstone and basalt.

2 Slate fine-grained, often gray, foliated rock split easily along cleavage planes of mica flakes aligned by pressure; formed from shale.

3 Marble Granular or sugary-textured rock; formed from limestone.

4 Quartzite Very hard, granular quartz rock; formed from sandstone.

5 Phyllite Silky, foliated rock more coarsely grained than slate, its usual precursor.

6 Schist Foliated rock, more coarsely grained and of higher metamorphic grade than phyllite; formed from slate or basalt.

7 Amphibolite Foliated rock of higher metamorphic grade than schist; formed from basalt.

8 Gneiss Foliated, banded, rock; coarser grained than schist and of the highest metamorphic grade.

Chapter 6

CRUMBLING ROCKS

As soon as rock is raised above the sea, the weather starts to break it up. Water, ice, and chemicals split, dissolve, or rot the rocky surface till it crumbles. Crumbled rock mixed with water, air, and plant and animal remains, forms soil. Soil and broken rock fall, flow, or creep downhill. This movement helps create the slopes that shape the surface of the land.

Chimney Rock on the Mackinac waterway, Michigan, and cliffs in the Yellowstone National Park, Wyoming. (Engravings from *Picturesque America* 1894)

Rocks attacked by weather 1

Wearing down of land begins as weather rots rocks at or near the surface. Different weather "weapons" probe different weaknesses – for instance the joints in igneous rocks like basalt and granite, bedding planes in clays and shales, and natural cements in sandstones and conglomerates. Chemical weathering (pp. 104-105) attacks chemical ingredients in rocks. Physical (mechanical) weathering destroys rock but leaves its chemicals unchanged. Which kind of weathering predominates depends upon the type of rock and climate.

Physical weathering is active chiefly in cold or dry climates. Agents include sharp temperature changes (reinforcing chemical effects), frost, drought, crystallizing salts, and growing plants.

1
Exfoliation
Cutaway view of a boulder
subject to exfoliation

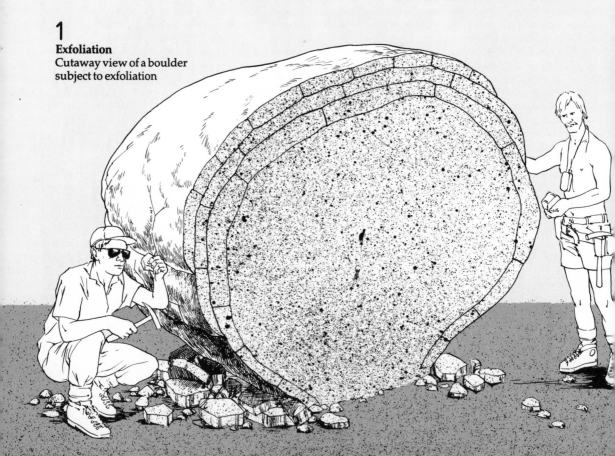

1 Exfoliation (spheroidal weathering) is the flaking of intensely heated surface rock as it expands more than the cooler rock below. This process produces rounded, isolated rock masses called exfoliation domes.

2 Block disintegration involves sharp temperature changes making desert rocks expand and contract. This helps enlarge the joints in rock, so splitting large masses into smaller blocks.

3 Frost action Water expanding as it freezes widens crevices in well-bedded or well-jointed rock and shatters it – in winter in mid-latitudes, at night in high mountains everywhere. Products are piles of sharp-edged debris including cone-shaped slopes of talus (scree) seen below steep peaks. Freezing also causes the granular disintegration of porous rocks like chalk.

4 Tree roots can widen cracks in rock as they grow.

5 Pressure release (unloading) follows removal of overlying rock and its pressure on the rock below. Expansion of that rock then forms curved joints promoting sheeting (pulling off) of rock shells from the inner mass – the likely origin of the huge domes in California's Yosemite Valley.

6 Slaking is the crumbling of clay-rich sedimentary rocks as they dry out during drought.

7 Crystallization of salts Dissolved salts expanding as they dry and crystallize in rock split the rock and honeycomb its surface.

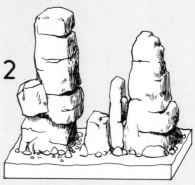

Block disintegration

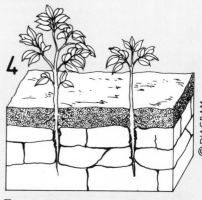

Frost action

Tree root action

© DIAGRAM

5 **Pressure release**
Yosemite Valley domes

Rocks attacked by weather 2

Chemical weathering attacks rocks aggressively in humid climates. The chief destructive agents are rainwater and certain substances that it contains. These dissolve some kinds of rock or rot the natural cements that bind particles in rock together. But not all minerals are equally at risk. Quartz proves much more resistant to attack than augite, biotite, hornblende, or orthoclase.

The following items summarize important ways in which chemicals make rocks decay.

1 Carbonation is the dissolving of limy rocks by percolating rainwater armed with carbon dioxide from the atmosphere or soil. The resulting weak carbonic acid widens joints in carboniferous limestone surfaces, producing bare limestone pavements where clints (sharp ridges) alternate with grikes (solution grooves). With items such as caves, swallowholes, and gorges these features form karst landscapes, named from a limestone region in Yugoslavia (see also pp. 128-129).

2 Hydration is when some minerals take up water and expand, breaking shells from the rock containing them. Sub-surface hydration probably produced southwest England's great rounded granite moorland blocks called tors.

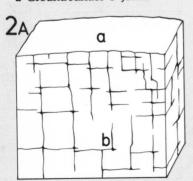

1 Carbonation
a Limestone pavement
b Clints
c Grikes
d Joints
e Bedding planes

2 Hydration
Three likely stages in tor formation:
A Unweathered granite mass
a Ground surface b Joints

B Weathering in process
c Joints widened
d Granite mass split into blocks between joints

C Tor bared by erosion of weathered material
e Massive blocks exposed
f Lowered ground surface

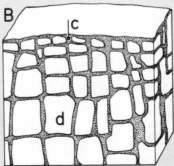

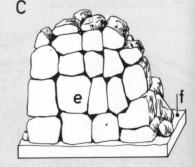

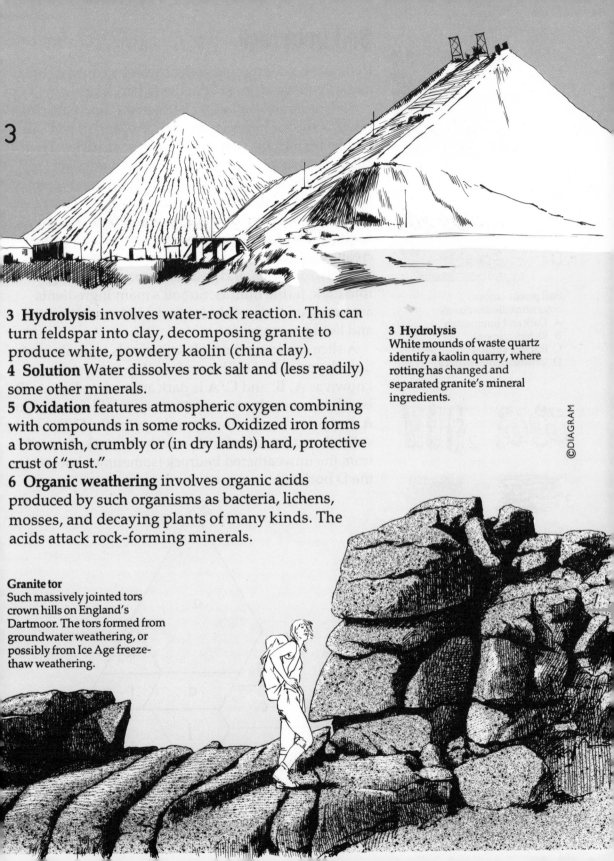

3 Hydrolysis involves water-rock reaction. This can turn feldspar into clay, decomposing granite to produce white, powdery kaolin (china clay).

4 Solution Water dissolves rock salt and (less readily) some other minerals.

5 Oxidation features atmospheric oxygen combining with compounds in some rocks. Oxidized iron forms a brownish, crumbly or (in dry lands) hard, protective crust of "rust."

6 Organic weathering involves organic acids produced by such organisms as bacteria, lichens, mosses, and decaying plants of many kinds. The acids attack rock-forming minerals.

3 Hydrolysis
White mounds of waste quartz identify a kaolin quarry, where rotting has changed and separated granite's mineral ingredients.

©DIAGRAM

Granite tor
Such massively jointed tors crown hills on England's Dartmoor. The tors formed from groundwater weathering, or possibly from Ice Age freeze-thaw weathering.

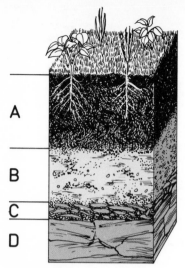

Soil profile (above)
Four simplified horizons:
A Dark and humus rich
B Rich in minerals
C Infertile subsoil
D Unweathered bedrock

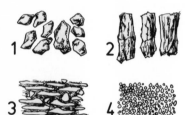

Soil structure (above)
1 Blocky 2 Prismatic
3 Platy 4 Crumb

Textural classes (right)
Soil texture varies with particle
size from clay (fine) through silt
(medium) to sand (coarse). Water
drains fast through sand, but clay
gets waterlogged. Loam is best for
plant growth.
a Clay b Sandy clay
c Sandy clay loam d Clay loam
e Silty clay
f Silty clay loam
g Sand h Loamy sand
i Sandy loam j Loam
k Silt loam l Silt

Soil from rock

In time most weathered rock acquires a covering of
soil – a substance most land life depends on.

Soil forms as weathering breaks rock into particles
ranging in size from clay to silt, sand, and gravel. Air
and water fill gaps between the larger particles. Then
chemical changes help bacteria, fungi, and plants to
move in. Plant roots bind groups of particles together;
leaves ward off destructive rain; roots and stems raise
minerals from deep down. Plants and their remains
form food for burrowing insects, worms, and larger
creatures. Bacteria and fungi decompose dead plants,
animal droppings, and dead animals, converting all
into dark, fertile humus. So soil's main ingredients
are (inorganic) minerals, (organic) humus, air, water,
and living organisms.

A slice cut down through soil reveals a profile made
up of layers called horizons, from top to bottom
known as A, B, and C. A is dark and rich in humus. B
is rich in minerals and substances washed down from
A, but paler, more compact and less fertile. C, the
subsoil, consists of infertile weathered rock, derived
from the unweathered bedrock (sometimes known as
the D horizon).

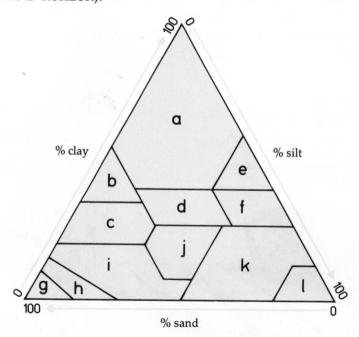

Few soils fit this description perfectly. Soil texture depends largely on the bedrock (parent rock), and soil type upon topography, time, vegetation and climate. Thus shales yield finer textured soils than sandstones. Soils on limestones are rich in bases; others have an acid tendency. Soil depth can range from less than an inch (2.5 cm) on steep slopes to several yards (m) on plains. Hillside soils are often better drained than those in valleys. Plants with different needs affect proportions of some substances accumulating in the soil. But the chief influence on soils is climate (see pp. 108–109).

Life in the soil (below)
These organisms live in or help to form the soil.
A Larger "aerators"
a Mole
b Earthworm
B Microorganisms
c Fungus
d Alga **e** Virus
f Bacterium
g Protozoan
C Arthropods
h Woodlouse
i Millipede
j Springtail
k Cockchafer larva
l Cricket
m Ant **n** Mite

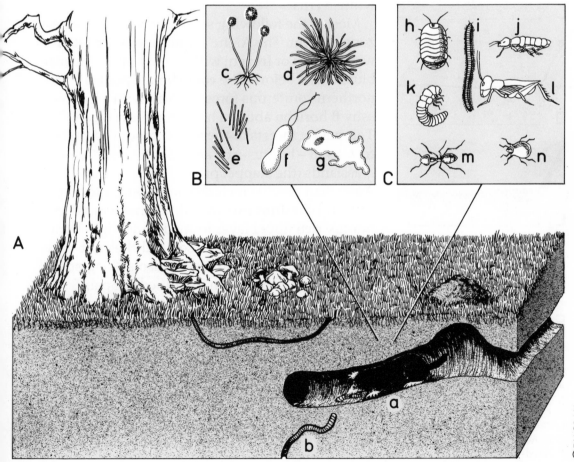

©DIAGRAM

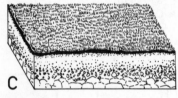

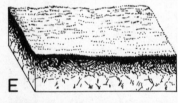

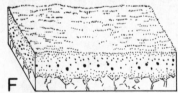

Six soil types
Diagrams above show six soil types.
A Tundra soil
B Desert soil
C Chernozem
D Ferralsol
E Brown forest soil
F A red-yellow podzol

Types of soil

Pedologists (soil scientists) have many ways of classifying soils. Thus the US Department of Agriculture labels soil types purely according to their properties, but the much-used zonal system stresses climatic origins.

Soils certainly owe more to climate and vegetation than to bedrock. Heavy rainfall causes much leaching (downward flow of dissolved substances), eluviation (downward movement of fine particles), and illuviation (redeposition of these substances at lower levels), and so affects soil fertility. At the other extreme hot deserts undergo salinization or alkalization as soil water brings salts and alkalis to the surface, then evaporates leaving them to form a whitish crust.

Most of the following soil types come from the zonal system. The Arctic's **tundra soils** are often waterlogged or frozen, with a peaty upper layer and bluish mud below. The **cool-climate podzol** ("ash") of northern coniferous forests is acid with a leached, ashy **B** horizon above a hard thin illuvial pan. Temperate forests of the world produced **brown forest soil** – humus-rich and slightly acid. Temperate grasslands (the steppe, prairie, pampas and Australian downs) include **chernozem**, and/or **prairie soil**, and **chestnut-brown** soil. Chernozem's dark, humus-rich upper layer formed under light rainfall with little leaching. Prairie soil and chestnut-brown soil formed in drier climates. **Alfisols** ("degraded chernozems") have been identified in Spain, northwest Africa, India, and parts of Africa. Hot deserts have pale, coarse, soils, poor in humus and sometimes white with salty crust. Tropical grasslands include dark, clayey **grumusols**. Deep, reddish, iron-rich **ferralsols** underlie the humid tropics. **Mountain soils** include thin scree soils – little more than rock fragments.

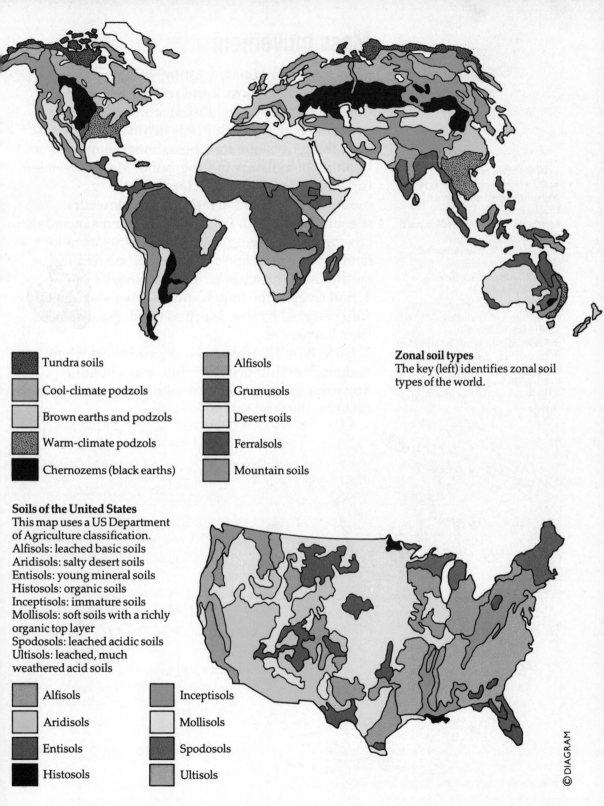

Zonal soil types
The key (left) identifies zonal soil types of the world.

Key (left):
- Tundra soils
- Cool-climate podzols
- Brown earths and podzols
- Warm-climate podzols
- Chernozems (black earths)
- Alfisols
- Grumusols
- Desert soils
- Ferralsols
- Mountain soils

Soils of the United States
This map uses a US Department of Agriculture classification.
Alfisols: leached basic soils
Aridisols: salty desert soils
Entisols: young mineral soils
Histosols: organic soils
Inceptisols: immature soils
Mollisols: soft soils with a richly organic top layer
Spodosols: leached acidic soils
Ultisols: leached, much weathered acid soils

- Alfisols
- Aridisols
- Entisols
- Histosols
- Inceptisols
- Mollisols
- Spodosols
- Ultisols

©DIAGRAM

109

Mass movement

Mass movement (mass wasting) is the force of gravity shifting weathered rock and soil (the regolith) downhill. Water often lubricates and aids this process. Amounts and speeds involved vary with such things as slope steepness, underlying rocks, and quantity of moisture in the ground. Mass movement can affect a few square yards or a whole mountainside. Sodden regolith may flow; dry regolith will slide or fall. Slow movements include creeping and flowing. Swift movements include landslides and rockfalls triggered by such things as earthquakes, heavy rain, or quarrying.

1 Soil creep is the imperceptible downslope creep of soil, betrayed by items such as tilted and displaced trees and fences.

2 Earth flow This features a stepped slope where material has slumped downhill, and a bulging downslope "toe" where it accumulates. Earth flow can occur in hours on saturated slopes.

Types of mass movement
Numbered block diagrams tally with numbered text items.
1 Soil creep with (**a**) bent weathered rock and (**b**) displaced trees, rocks, etc
2 Earth flow with (**c**) stepped slope and (**d**) downslope "toe"
3 Mudflows issuing from canyons in a desert
4 Slumping with (**e**) sandstone slump block after sliding down (**f**) a cliff of weak shale
5 Rockslide showing (**g**) scar torn in mountainside and (**h**) landslide debris
6 Rockfalls producing (**i**) scree or talus cones of debris fallen from a cliff (**j**)

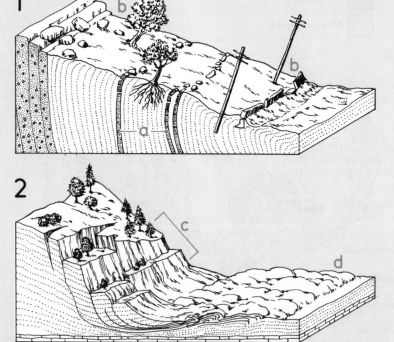

3 Mudflows occur where mud holds so much water that it flows as slurry even down a gentle slope. A mudflow containing volcanic ash and dust from Mt. Vesuvius swallowed the Roman city of Herculaneum in AD 79.

4 Slumping is a landslide where rock masses tilt back as they slide from a cliff or escarpment – often where well-jointed sedimentary rocks overlie clay or shale. Resulting slump blocks can be 2mi (3km) long and 500ft (150m) thick.

5 Rockslides involve masses of bedrock slipping down a sloping fault or bedding plane. Rockslides have killed people or destroyed villages in the Canadian Rockies, Norway, and Switzerland.

6 Rockfalls are free falls of fragments of any size from a cliff. Frost-shattered fragments in time form cliff-foot talus cones with a talus (or scree) slope at an angle of about 35 degrees.

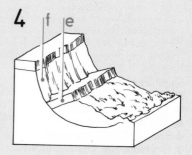

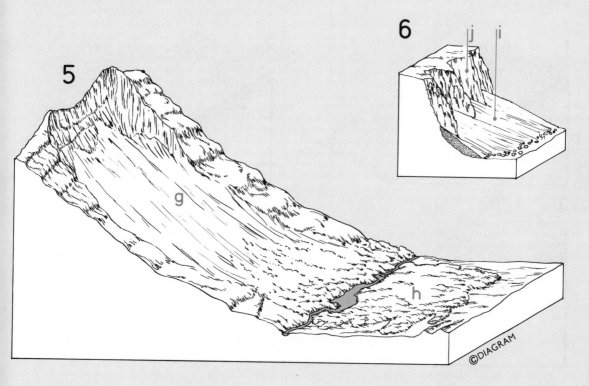

©DIAGRAM

Slopes

Slopes steep or gentle make up almost all the surface of the land. Everywhere, mass movement, splashing raindrops, or rainwater flowing over land are forming slopes and wearing them away by shifting soil or broken bits of rock downhill.

Wherever the force of gravity is greater than the force of friction holding particles upon a slope these tend to slide downhill. Most slopes have an average angle of less than 45 degrees. But a single slope usually has several (straight, concave, or convex) segments – parts with different angles. Slope angle varies with the amount of weathered debris entering and leaving a segment.

Slope segment angles
Block diagrams show slope steepness influenced by input and output of (**a**) weathered debris overlying (**b**) bedrock.
1 Graded slope: Debris output equals input with none added by the segment.

2 Graded slope: Output equals input including debris added by the segment.
3 Steep slope: Input exceeds output.
4 Gentle slope: Output exceeds input.

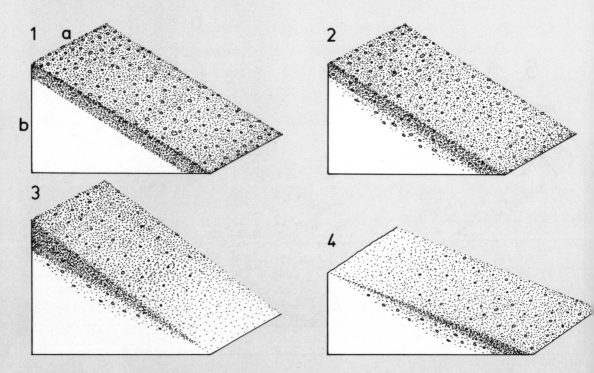

Slopes probably evolve in three main ways, all possible in one locality:

A Slope decline Slope angle decreases through stages: (**a**) steep free face; (**b**) graded slope with convex curve above, concave curve below; (**c**) decreasing curvature; (**d**) reduction in height. Slope decline predominates in moist temperate areas such as the northeast US Appalachians.

B Slope retreat The retreating slope keeps a short convex top, long free face, debris slope, and (lengthening) pediment (thin sheet of debris). Such slopes abound in semiarid areas.

C Slope replacement Lower angle slopes extend upward to replace steep upper segments.

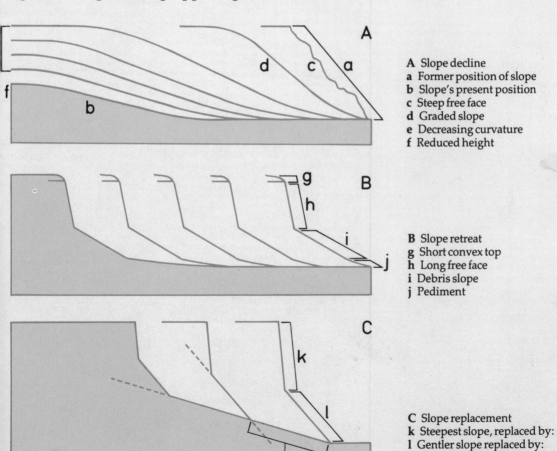

A Slope decline
a Former position of slope
b Slope's present position
c Steep free face
d Graded slope
e Decreasing curvature
f Reduced height

B Slope retreat
g Short convex top
h Long free face
i Debris slope
j Pediment

C Slope replacement
k Steepest slope, replaced by:
l Gentler slope replaced by:
m Gentlest slope

©DIAGRAM

Chapter 7

HOW RIVERS SHAPE THE LAND

Rain-fed rivers carve valleys in the hills, wash soil and weathered rock downhill, and dump them in lakes and seas. Year by year the erosive work of rivers wears down the highest mountain ranges until some form flat plains that barely peep above the sea. Off some sheltered shores, though, sediments shed by rivers build deltas – muddy aprons of new land. This chapter describes the work of rivers, and the landforms many help create.

The Yellowstone River Valley, Wyoming. (Engraving from *Picturesque America* 1894)

Running water

Soon after land appears above the sea, rivers set about attacking it. Rivers rise in highlands and flow downhill to empty in a sea or inland drainage basin. The force of their moving water erodes and transports a load of soil and rock, so carving valleys that dissect mountains into peaks and ridges and reducing these to hills. But rivers also deposit the eroded debris to build lowland plains and offshore underwater platforms.

In theory, by degrading (eroding) some stretches of its bed and aggrading (building) others, a river tends to gain a graded concave profile, leaving it with just enough velocity to shift its load. In theory, too, rivers tend to bevel continents to peneplains – almost level plains just above sea level. In practice, earth movements and differences in rock formation and resistance to erosion interrupt both trends.

But calculations confirm running water's power to sculpt the continents. Rivers drain about 70 per cent of all dry land. They contain only 0.03 per cent of all fresh water, yet carry enough each year to drown all dry land 1ft (30cm) deep. Every year rivers dump about 20 billion tons of eroded material in the sea – enough to shave 1.2in (3.13cm) off the Earth's land surface every thousand years. The Mississippi River alone shifts an estimated 516 million tons a year: 340 million tons as fine suspended particles, 136 million tons in solution, and 40 million tons by saltation – the hopping of heavy particles along the river bed.

An individual river's power to erode and transport depends largely on its discharge – the product of its volume and velocity. The greatest mean discharge is the Amazon's 6,350,000 cu ft per sec (180,000 cu m per sec) – nearly ten times the Mississippi's. But for rivers such as China's Yangtse, wet and dry seasons enormously affect the flow from month to month.

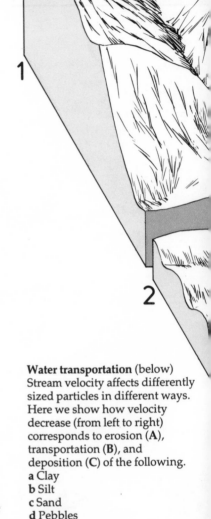

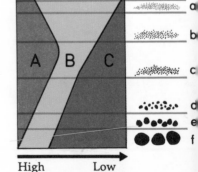

Water transportation (below) Stream velocity affects differently sized particles in different ways. Here we show how velocity decrease (from left to right) corresponds to erosion (**A**), transportation (**B**), and deposition (**C**) of the following.
a Clay
b Silt
c Sand
d Pebbles
e Cobbles
f Boulders

High velocity   Low velocity

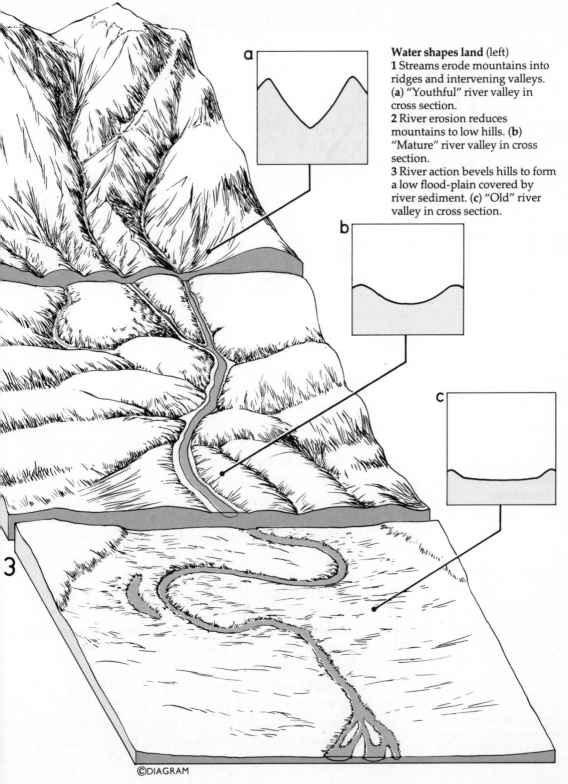

Water shapes land (left)
1 Streams erode mountains into ridges and intervening valleys. (**a**) "Youthful" river valley in cross section.
2 River erosion reduces mountains to low hills. (**b**) "Mature" river valley in cross section.
3 River action bevels hills to form a low flood-plain covered by river sediment. (**c**) "Old" river valley in cross section.

a

b

c

3

©DIAGRAM

Water comes and goes

The water cycle (above)
Arrows indicate the circulation of moisture that keeps rivers running.
a Sea
b Evaporation
c Cloud formation
d Wind
e Precipitation
f Rivers

1

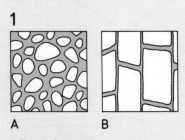

A   B

1 Ground water
A Water percolates between the grains of sand in sandstone, a porous permeable rock.
B Water only percolates through joints in limestone, a non-porous permeable rock.

The chief agent wearing down dry land is water flowing on the surface as part of the water cycle. Powered by the Sun and gravity, the cycle starts as the Sun's heat evaporates surface water, mostly from the oceans. Water vapor in the atmosphere cools and condenses into droplets that build clouds. Clouds shed this moisture as rain or snow. Much of this precipitation falls on the sea, but some falls on dry land. Vast quantities are locked in slowly moving sheets of ice (see pp. 154–155). But some water seeps underground, and smaller quantities run off the surface as rills, converging in rivers and lakes. Most water returns to the sea.

Rainfall is occasional but many rivers are perennial, fed by underground water. This includes the following features.

1 Ground water Below the soil ground water saturates permeable rocks (rocks that let water through), filling the pores of porous rocks such as sandstone, and cracks in pervious rocks including limestone.

118

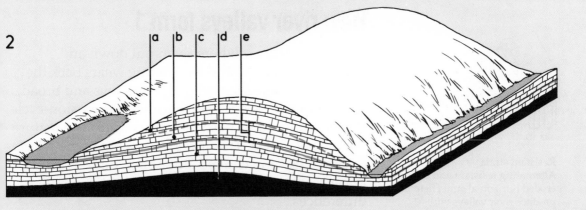

2

2 Aquifer This is a saturated layer of permeable rock lying on a layer of impermeable rock such as slate or shale. The aquifer's surface is its water table. The table's level varies with rainfall, tends to follow slopes up and down, and is exposed in swamps, lakes, and springs.

3 Spring This is a flow of water escaping from the ground, as where the water table outcrops on a hillside above impermeable rock. Springs are the sources of some major rivers.

4 Artesian basin This is a saucer-shaped aquifer sandwiched between layers of impermeable rock. Rain soaks down through its rim. Below rim level water under pressure may gush from so-called artesian wells and artesian springs. Artesian basins underlie London, Paris, and much of the Sahara Desert, Australia, and North America.

2 Aquifer
a Non-saturated rock
b Rock sometimes saturated
c Rock always saturated
d Impermeable rock
e (Variable) water table

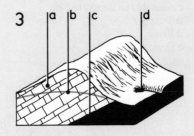

3

3 Spring
a Porous rock
b Water table
c Impermeable rock
d Spring

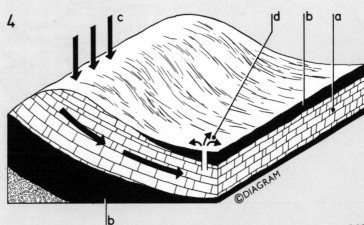

4

4 Artesian basin
a Permeable rock (aquifer)
b Impermeable rock
c Rainfall
d Artesian well

©DIAGRAM

How river valleys form 1

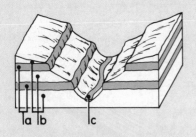

Resistant strata
Alternating resistant and easily eroded horizontal strata help to produce river valleys with a stepped cross-profile.
a Resistant strata
b Easily eroded strata
c River

A mountain valley
a Gully
b Tributary stream
c Alluvial fan deposited by tributary
d River
e Divide (see p. 130)

River valleys grow where rivers cut down and sideways into rock, and weathering wears back the slopes on either side. Steep-sided valleys and broad, flat-bottomed valleys were once seen as erosion-cycle stages produced respectively by youthful and mature sections of a river. This takes no account of the effects of local rocks and climate. But upper, middle, and lower reaches of a river do often show distinctive differences.

Many valleys start to form high up on slopes where springs erupt or rainwater successively produces splash, sheet, then rill erosion – for deepened rills cut river channels. In its upper course a river may be a small, fast-flowing torrent cutting down into its bed, and forming rapids and waterfalls (see also pp. 122–123.) Floods accelerate this process, transporting rocks that gouge out rock pools. Mountain streams commonly reveal these features:

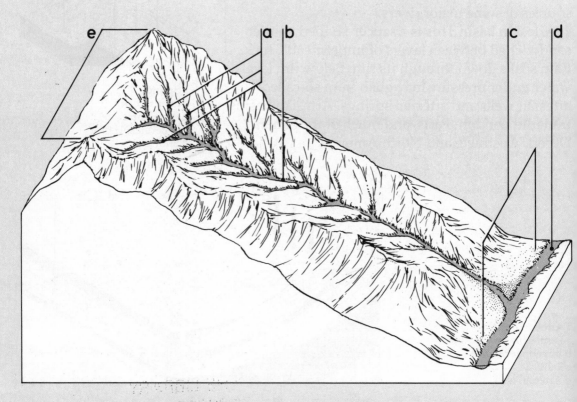

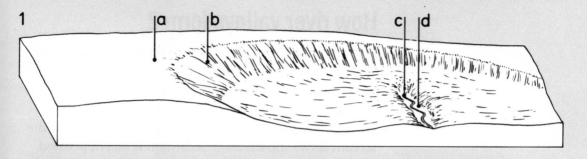

1 Headward erosion Undercutting, rain wash, and soil creep help a river valley to gnaw back into a hillside, so lengthening the river.

2 Pot-holes Circular holes in a rocky stream bed show where waterborne scraps of rock were whirled around, deepening depressions. Stream-channel erosion by rock fragments is called corrasion. Erosion also rubs bits off the rock fragments themselves, a process called attrition.

3 V-shaped valley Valleys develop deep, steep cross-sections where streams erode downwards. Interlocking spurs are ridges projecting from both sides of the valley. The stream erodes most on the outsides of bends, where the current flow is strongest, so the zigzags grow more pronounced.

1 Headward erosion
a Rainwash
b Undercutting and soil creep
c Source of river
d Upper river valley

2 Pot-hole
a Stream flow
b Path of pebble
c Stream bed
d Pot-hole

3 V-shaped valley
a V-shaped cross section
b Interlocking spurs
c Stream

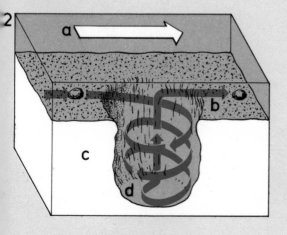

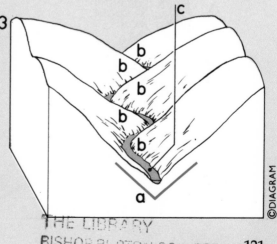

©DIAGRAM

How river valleys form 2

A river valley's middle section typically shows more mature features than its upper course. Weathering has broadened the valley sides. The river flows down a gentler gradient. Its current travels fast enough to shift the load of sediment acquired upstream. But the stream flows mostly over sediment it has deposited, and no longer cuts down into the rock below its bed. Rather, it erodes from side to side, nibbling away its banks, so flattening and broadening the valley floor. Yet even in this middle section, land uplift or an outcrop of resistant rock may interrupt the valley's gently curving profile with a waterfall.

Illustrations show these processes at work.

1 Meander This is where a river current flows around a bend, hits its concave bank at speed, and gnaws that bank away to form a river cliff. Meanwhile, the river deposits sediment in slack water on the convex bend, where a sloping spur known as a slip-off slope grows out into the river. So the river migrates slowly sideways.

1 Meander
This diagram of a meander (marked curve in a river channel) shows these features.
a Erosive surface flow
b Concave bank
c River cliff
d Flow on river bed, shedding sediment
e Convex bank
f Shingle
g Slip-off slope

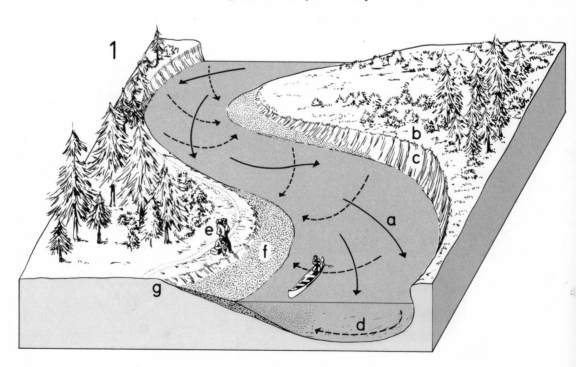

2

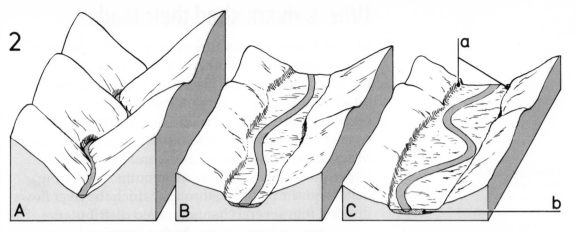

A B C a b

2 Valley broadening Meanders slowly migrate downstream, widening the river valley by lopping off the ends of interlocking spurs. This rims the valley floor with low cliffs called bluffs.

3 Waterfall This is a cliff-like face down which a river plunges. Waterfalls form where resistant rock or land uplift interrupt the river's profile. Major falls include the Zaïre River's Boyoma (Stanley) Falls, the Paraná's Guaíra Falls, the Niagara Falls, and the Zambezi's Victoria Falls. Vertical river erosion cuts back some waterfalls so they migrate upstream above a long, deep gorge.

2 Valley broadening (above)
A sequence of three diagrams shows migrating meanders widening a valley floor.
A Lateral erosion starts at concave banks.
B Lateral erosion lops off spurs.
C Large meanders migrating downstream broaden the valley floor, carving bluffs (**a**) and dumping gravel (**b**).

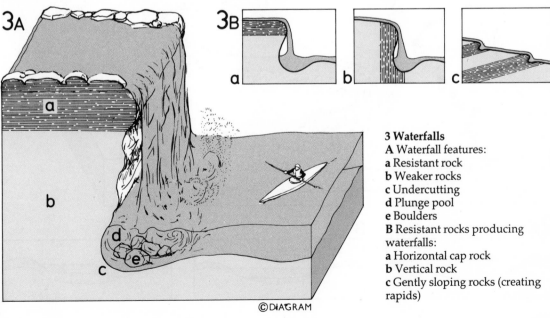

3A 3B a b c

a b d c e

3 Waterfalls
A Waterfall features:
a Resistant rock
b Weaker rocks
c Undercutting
d Plunge pool
e Boulders
B Resistant rocks producing waterfalls:
a Horizontal cap rock
b Vertical rock
c Gently sloping rocks (creating rapids)

©DIAGRAM

Where rivers shed their loads

In their lower courses, huge rivers like the Amazon and Mississippi flow down a gradient as slight as 3in per mile (5cm per km). Here, the rivers have beveled off all hills and cross a broad, low-lying plain thickly carpeted with sand and mud. Vast quantities of these river sediments built the flood-plains of the Ganges, Mississippi, Niger, Nile, Rhine, and Rhône. Deposition blocks such rivers' mouths, producing deltas – swampy plains through which the river flows divided into several channels called distributaries. Deltas grow where a river sheds a large load of sediment faster than tides and currents can carry it away.

Illustrations show flood-plain and delta features.

1 Flood plain This has a valley floor flattened and broadened by river erosion, and floored with sediments deposited by migrating meanders and river floods. Cut-off meanders survive as ox-bow lakes. The river flows above the level of the plain and between raised banks called levees.

1 Flood plain (below)
These features appear in a river's flood plain.
a Snaky meanders
b Levees
c Ox-bow lakes
d Mud, silt and sand
e Bluffs

2 Levees
Cross sections show how a river raises its banks.
A Sediment deposited by flooding
B Sediment deposited in normal flow
C Sediments and river after repeated flooding

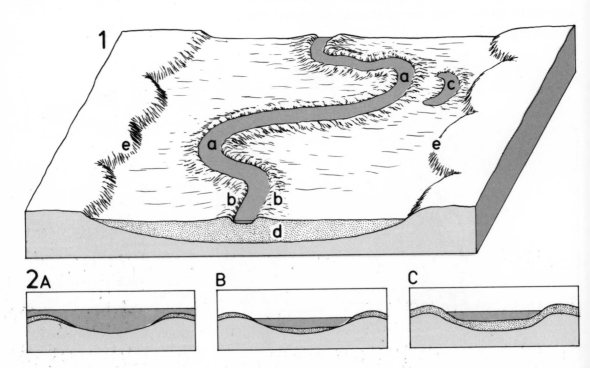

124

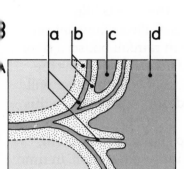

2 Levees form because the river floods repeatedly. A flooding river sheds mud on its banks where current flow is slow, so banks grow higher. After flooding, the river deposits sediment on its bed, so raising it. After repeated flooding both bed and banks are raised, and the river surface lies higher than the plain on either side.

3 Delta Deltas take the shape of a fan or bird's foot. They form in stages. Deposition splits a river into distributaries flanked by levees and separated by lagoons. Sediments turn lagoons into swamps and build bars and spits. Swamps become dry land (Louisiana, Mesopotamia, and the North China Plain derived from deltas in this way).

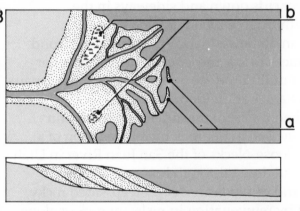

3 Deltas
Plans and longitudinal sections show how a delta forms.
A The river splits up.
a Distributaries
b Levees
c Lagoon
d Sea
B The delta extends into the sea.
a Spits and bars
b Swamps
C The delta expands.
a Infilled swamps become dry land.

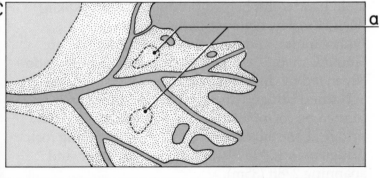

©DIAGRAM

125

Rivers revived

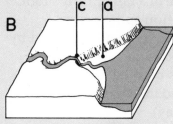

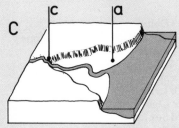

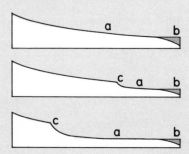

1 Knickpoint
Block diagrams (above) and long profiles (below) show a knickpoint receding upstream from a river mouth.
A Before land uplift
B Soon after uplift
C Later
a Flood plain
b Sea
c Knickpoint

Mature, meandering rivers sometimes start vigorously cutting down into their beds like "youthful" mountain torrents. Called rejuvenation, this happens if a lasting rise in rainfall boosts a river's flow, or if land uplift or a fall in sea level leaves the river mouth above the sea. Either way the river will regrade its bed. Various landforms show where this has happened.

1 Knickpoint This is a sharp step in a river's long profile, often marked by rapids or a waterfall. In time a knickpoint starting at a river's mouth recedes upstream. Some rivers show several knickpoints resulting from successive uplifts of the land.

2 River terraces are remnants of old flood plain left when a river cuts down and sideways in the sediments through which it flowed before rejuvenation. Renewed rejuvenation cuts a second pair of terraces below the level of the first. Further episodes of uplift produce more terraces. Several pairs flank the Thames at London.

3 Entrenched (incised) meanders are floodplain meanders incised in bedrock by rejuvenation of a river. The Goose-Necks of the San Juan River in Utah have been cut down through horizontal beds.

4 Gorges and canyons are deep, steep-sided rocky valleys cut by rejuvenation in resistant rock or along a fault. Land uplift helped produce the 3mi (4.8km) deep Himalayan gorges of the Ganges, and Brahmaputra, and Arizona's 1.5mi (2.4km)·deep Grand Canyon – a desert gorge cut by the Colorado River, fed by distant melting snows.

5 Natural bridge This type of rock formation is formed by cut-off of an incised meander. Rainbow Bridge at Navajo Mountain, Utah, is a striking rock bridge rising 309ft (94m) from a gorge floor, and spanning 278ft (85m).

2 River terraces
a Oldest terraces
b Younger terraces
c Present floodplain
d Present knickpoint
e Sediments

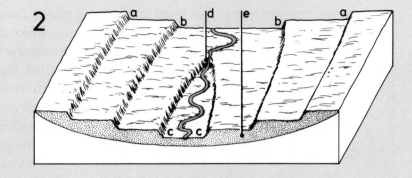

3 Entrenched meanders
River erosion has kept pace with uplift of a floodplain by deeply incising old meanders in the rising land.

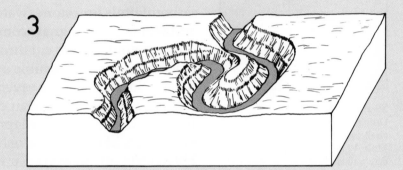

4 Gorges
River rejuvenation cut this long, deep, narrow gorge in land rising as earth movements built mountains.
a Mountain
b Gorge

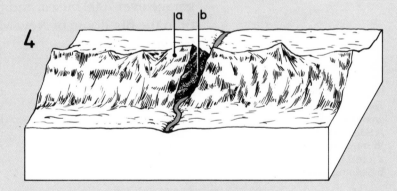

5 Natural bridge
Lateral river erosion whittled away cliffs flanked by an incised meander. Then the stream cut through the meander's narrow neck.
a Old incised meander
b Present course of river
c Natural bridge

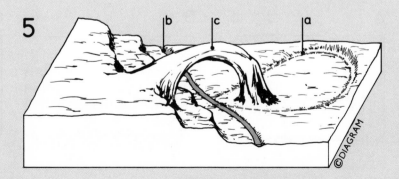

127

Rivers underground

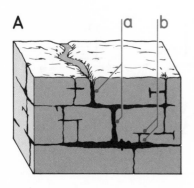

Swallowholes (above)
Dots show swallowholes in the chalk bed of England's River Mole. Water tends to vanish down these holes in dry weather.

In most landscapes, rivers flow across the surface of the land. But rainwater sinks down through joints in permeable limestone rocks, and invisibly attacks them underground. For carbon dioxide gas combines with falling rain to turn the water into weak carbonic acid, and carbonic acid dissolves calcium carbonate – the main ingredient of limestone. In moist chalk countryside this process removes an estimated 35 tons of rock a year from every acre and forms depressions called solution hollows. In carboniferous limestone, water enlarges vertical and horizontal joints to gnaw complex channels underground. A slice cut through some limestone mountains would resemble a giant slice of Gruyère cheese. Some limestone caverns and cave systems are immense. Several European cave systems descend 4000 ft (1219m) or more. Mapped passages in Kentucky's Mammoth Cave National Park exceed 230mi (370km). Borneo's Sarawak Chamber could garage over 7000 buses, and five soccer pitches would fit in the Big Room of New Mexico's Carlsbad Caverns.

How caves form
Three illustrations trace the growth of caves in limestone.
A Rainwater trickles down through crevices produced by:
a Joints
b Bedding planes

B Rainwater acting as a weak carbonic acid dissolves the rock it touches and removes the dissolved material. This widens vertical and horizontal crevices in limestone.

C Streams plunging underground widen crevices into vertical and horizontal caves. If the climate changes so that rainfall drops, many caves are left quite dry.

A

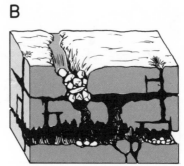

B

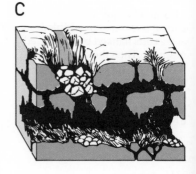

C

Where dripping water evaporates and/or gives up carbon dioxide the dissolved calcium bicarbonate becomes insoluble in the mineral form called calcite, building deposits such as these:

1 Stalactites "Icicles" of calcite hanging from a cave ceiling. According to conditions they take 4 or 4000 years to grow one inch (2.54cm).

2 Stalagmites Calcite spikes jutting upward from a cave floor.

3 Columns Calcite forms produced where stalactites and stalagmites meet.

4 Gours Calcite ridges formed where water rich in carbonate flows over an irregular surface.

A limestone cave system
a Clints (blocks)
b Grikes (gullies)
c Sink-holes
d Galleries
e Stalagmites
f Stalactites
g Columns
h Gours

© DIAGRAM

129

Drainage patterns

Land drained by a river and its tributaries comprises a drainage basin, also called a river basin, catchment area and (in the US) watershed – a term in Britain meaning a divide: the high ground between two drainage basins. Some drainage basins are immense. The Amazon's, the world's largest, embraces almost two-fifths of South America.

A drainage basin's river channels form a drainage pattern that depends on slope, rock types and formations, and crustal movements. Drainage patterns include the following types and features.

1 Dendritic pattern Named from the Greek *dendron*, a tree, this is a tree-shaped pattern of small, branching tributaries feeding into a main "trunk" river. It forms on rocks of equal resistance.

2 Trellis pattern Here some tributaries flow parallel with the main river, others at right angles to it. This pattern appears where bands of hard and soft rock alternate. (See also pp. 132–133.)

Parallel pattern (below)
Parallel tributaries join at an acute angle. This pattern is typical of young river systems on uniform rock.

1 Dendritic pattern
Small branching tributaries flowing on uniform rock unite to form a network like the twigs, branches, and main trunk of a tree.

2 Trellis pattern
a Consequent stream (following the land's original slope).
b Subsequent stream (at right angles to consequent stream).
c Obsequent stream (opposite in direction to a consequent).

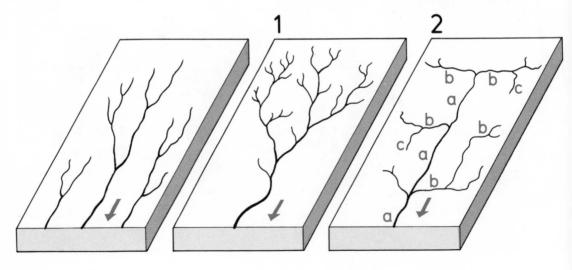

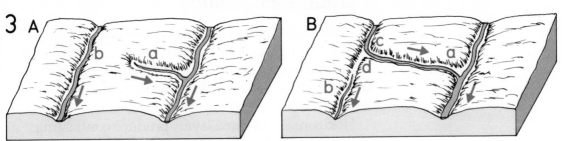

3 **River capture** This involves a stream eroding headward until it captures another stream's headwaters at a so-called elbow of capture. The "pirate" stream's flow increases, and the captured stream's flow dwindles. The victim stream may then rise below the elbow of capture, leaving a dry tract of valley called a wind gap.

4 **Accordant drainage** is where river channels relate to rock type and structure. On folded rocks, tributaries quickly attack tension cracks in (upcurved) anticlinal crests and often wear these down below intervening valleys formed in (downcurved) synclines. Synclinal crests occur in the Appalachian Ridge-and-Valley section. (Discordant drainage patterns show no relation to today's rock structure. They may be antecedent – formed before the land was raised or tilted – or superimposed on present rocks after developing on others that have worn away.)

3 River capture
A Headward erosion of (**a**) toward (**b**).
B (**a**) has captured (**b**)'s headwaters.
c Elbow of capture
d Wind gap

4 Accordant drainage
A Rivers flow off anticlinal slopes and along synclinal troughs.
a Anticline
b Syncline
B River erosion has lowered anticlines to valleys and left synclines upstanding as mountains.

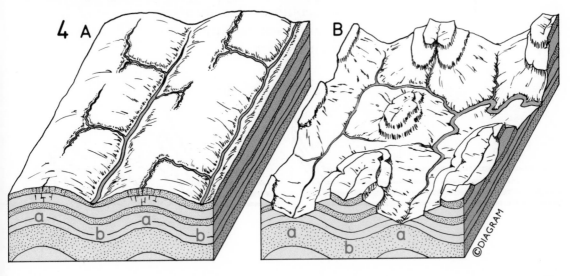

Plateaus and ridges

As valleys eat into sedimentary rocks they sculpt distinctive upland features determined largely by the types and angles of rock layer – especially where weak layers such as clay or shale alternate with more resistant layers like chalk, limestone, or sandstone.

Horizontal and tilted layers produce contrasting types of upland.

1 Plateaus develop where resistant rock caps other horizontal layers. River valleys may dissect a plateau into tablelands such as Brazil's *tableiros*, or the Colorado Plateau's steep-sided blocks called mesas, many now eroded into the smaller blocks called buttes (see also pp. 168-169). Plateaus may have "stepped" sides shaped by different rates of erosion of alternating weak and resistant rock layers.

2 Cuestas are ridges formed by gently tilted strata. Each has a steep slope or escarpment and a gentle slope or dip slope. Beyond the cuesta lies lower land where erosion has eaten deeply into weaker rock. Cuestas abound in the US Southwest, and occur along the Gulf and Atlantic coasts. Cuestas overlooking clay vales dominate southeast England's landscape, and rim part of the Paris Basin.

3 Hogbacks are steep, even-crested ridges which are formed by sharply dipping strata. They abound in the mid and southern Rockies and form most of the long, parallel ridges of the Appalachian Mountains' Ridge-and-Valley system.

1 Plateaus
"Before and after" block diagrams show the origins of buttes and mesas.
A Horizontal rock layers laid down beneath the sea
a Resistant rock
b Easily eroded rock
c Sea
B The same rock layers after uplift and dissection by downcutting rivers
d Mesa
e Butte
f River valleys

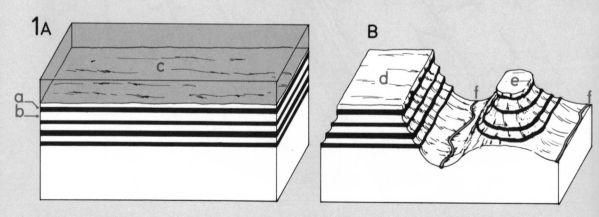

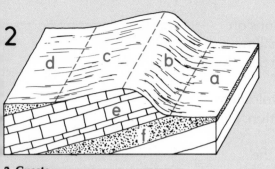

2 Cuesta
a Scarp foot
b Escarpment
c Dip slope
d Back slope
e Resistant rock layer
f Easily eroded layer

3 Hogback
a Steep slopes
b Steeply dipping bed of resistant rock
c Easily eroded rock bed

Cuestas and vales
Block diagrams (below) show dissection of a dome and basin of sedimentary rock layers into cuestas with intervening vales.
A Dissected dome (dip slopes facing outward)

B Dissected basin (dip slopes facing inward)
a Resistant rock layers
b Easily eroded rock layers
c Cuestas
d Vales

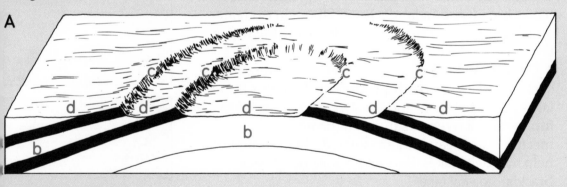

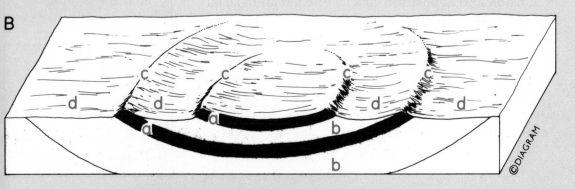

How lakes form

Lakes are bodies of water lying in depressions on land. They can be large or small, deep or shallow, fresh or salt. The largest lake is the Caspian Sea; the deepest, Lake Baikal in eastern Siberia.

Lakes are formed by earth movements, volcanoes, erosion, deposition, or erosion and deposition. Here are examples of lakes formed by each.

1 Earth movements Crustal uplift isolated the Caspian Sea from the Black Sea. Crustal warping (and later glacial action) formed Lake Superior. Fault blocks sinking between high valley walls produced the long, narrow trench containing Lake Malawi and other African Rift Valley lakes.

2 Volcanic action Water-filled volcanic craters include Oregon's Crater Lake. Lava flows damming river valleys penned back lakes like the Sea of Galilee and Lac d'Aydat in south central France.

3 Erosion lakes fill ice-worn hollows, especially in Canada and Finland. Water dissolving limestone river banks broadened Ireland's River Shannon to create Lough Derg. Wind eroding rock to below the water table exposes lakes in deserts.

4 Deposition Rock slides damming rivers form lakes such as Montana's Earthquake Lake. Ice dams trap lakes against glaciers and ice sheets. Ox-bow lakes form where a river cuts through the necks of meanders and leaves them isolated. River sediments help trap delta lakes. Bars and dunes pond back brackish coastal lakes.

5 Erosion and deposition Glacial erosion and glacial deposits called moraines formed countless lakes. Moraine-dammed cirques (ice-eroded hollows) hold circular mountain lakes called tarns. End moraines at the mouths of glaciated valleys dam long narrow lakes such as New York's Finger Lakes, England's Lake District lakes, Sweden's Glint-Line lakes, and Italy's lakes Como, Garda, and Maggiore.

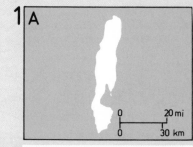

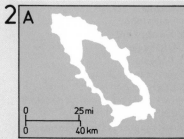

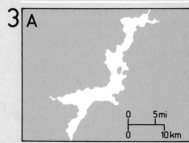

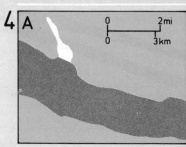

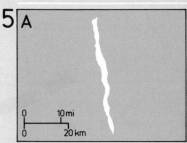

1 "Earth-movement" lakes
A Map of the Dead Sea, a salt lake in a rift valley
B Diagram of a deep rift-valley lake, formed in a crack in the Earth's crust

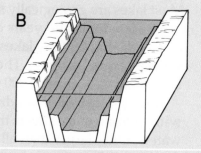

2 "Volcanic" lakes
A Map of Lake Toba, Sumatra, one of the world's largest volcanic crater lakes
B Diagram of a lake occupying a volanic crater
C Diagram of a lake dammed by a lava flow

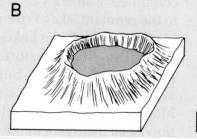

3 "Erosion" lakes
A Map of Lough Derg, Ireland
B Diagram of lakes in rock basins gouged from the rock by passing ice sheets
C Diagram of a lake in a deflation hollow where desert wind exposed the water table

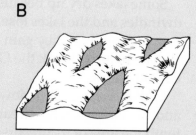

4 "Deposition" lakes
A Map of Lake Vatnsdalur, Iceland, dammed by ice
B Diagram of a river about to cut through a meander
C Diagram of an ox-bow lake – a meander cut off from the river by sediment

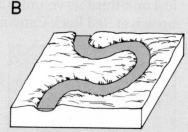

5 Lakes formed by erosion and deposition
A Map of Lake Seneca, one of New York's Finger Lakes
B Diagram of a moraine-dammed cirque
C Diagram of a moraine-dammed finger lake

©DIAGRAM

Vanishing lakes

Most lakes are geologically short-lived: they dry up in a mere few thousand years or so. A lake can disappear in several ways. Many lakes get clogged by mud and silt washed in by rivers. The River Rhône will fill in Lake Geneva within 40,000 years or so.

Huge prehistoric lakes drained away or shrank with melting of ice sheets that had ponded back their waters. Once larger than all today's Great Lakes combined, Canada's Lake Agassiz has been reduced to the remnant lakes Winnipeg, Winnipegosis, and Manitoba. The Great Lakes themselves are shrunken relics of mighty prehistoric Lake Algonquin. And about 15,000 years ago, bursting of an ice dam abruptly emptied Montana's mighty prehistoric Lake Missoula in a flood some 10 times greater than the flow of all the rivers in the world.

Some lakes dry up because the local rainfall dwindles and the lakes lose more water by evaporation than they gain from rivers. A drying climate helped shrink the Great Basin lakes of the United States.

Outlet rivers cutting down through bedrock are another factor. Utah's Great Salt Lake covers one-tenth the area of its precursor, Lake Bonneville, which lost one-third its volume in six weeks or so through a breach at Red Rock Canyon.

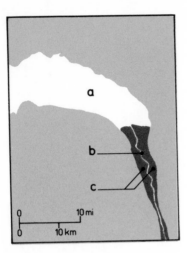

Lake Geneva (above)
A map shows how river sediment has filled in one end of this lake.
a Lake Geneva
b River Rhône
c Former lake, now floored by river sediment

Lost lakes of America (right)
Maps show lakes lost in relatively recent times, with their major modern relics.
A North-Central North America
a Former Lake Agassiz
b Lake Winnipeg
c Lake Winnipegosis
d Lake Manitoba
e Lake of the Woods
B Great Basin lakes
f Former Lake Lahontan
g Former Lake Bonneville
h Great Salt Lake

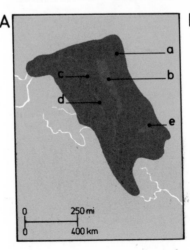

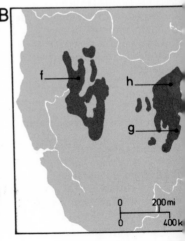

Disappearing lakes and ponds produce many of the world's wetlands. Swamps are places that are always waterlogged – for instance the Florida Everglades, Virginia's Dismal Swamp, and much of the Sudd in the Sudan. Marshes are low-lying lands that flood when rivers overflow. Bogs are soft, wet, spongy areas where moss fills shallow lakes and pools.

When even wetlands dry out, clues to vanished lakes remain in flat valley floors thickly carpeted with mud and silt, old overflow channels, and old beaches high on valley slopes.

A lake vanishes (below)
Diagrams show how a river fills in a lake.
A River washes sediment into one end of the lake.
B River sediment builds a delta out into the lake.
C Sediment shrinks the lake into a small, shallow, reedy swamp.
D The lake floor is now filled in with sediment colonized by land plants.

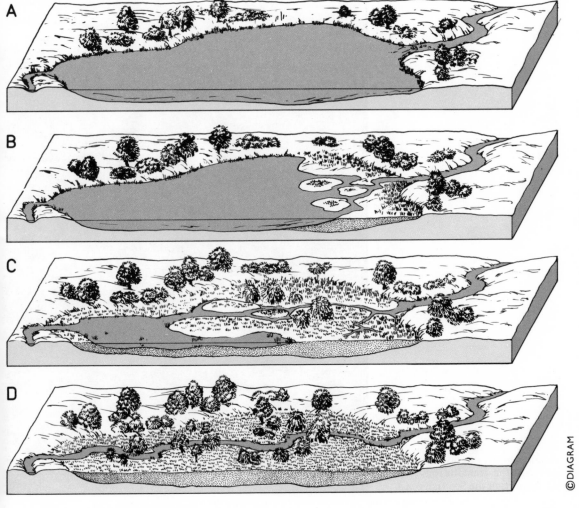

© DIAGRAM

137

Chapter 8

THE WORK OF THE SEA

Sea water set in violent motion scours coasts around the world. On rocky shores, waves armed with stones are battering the land away. Yet fragments torn from cliff-fringed coasts lodge on gently shelving shores to raise new sand and shingle beaches. This chapter shows how sea erodes and builds land. It describes, too, shores risen or submerged by changing levels of the sea or land. The chapter ends with the growth of coral reefs and islands.

A 19th century engraving showing the dramatic rock formations of the Scottish coast.

138

The sea in action

Besides shaping inland surfaces, water sculpts the coast – the zone where land meets sea. Coasts include sea cliffs, shores (areas between low water and the highest storm waves) and beaches (shore deposits). Sea water set in motion erodes cliffs, transports eroded debris along shores, and dumps it on beaches. So most coasts retreat or advance. The chief agents in this work are waves and currents, but tides contribute, too.

1 Waves are undulations mainly set in motion by the wind. Wave height and power depends upon wind strength and fetch – the amount of unobstructed ocean over which the wind has blown. In the open sea waves pass through the water without moving it forward. But in shallow water, wave crests crowd closer, pile up, and overbalance, causing forward movement of the water. The consequences are featured later in this chapter.

1 Waves
Waves trigger circling of water particles in a stack about half a wavelength deep. In water shallower than that depth the circles are flattened by the proximity of the bottom. The waves grow in height, become unstable, and their crests topple and break.
a Wave crest
b Wave trough
c Wave height
d Wave length

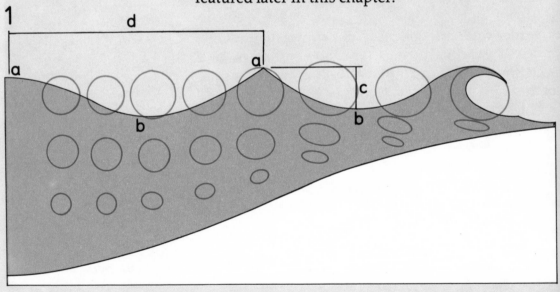

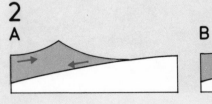

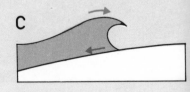

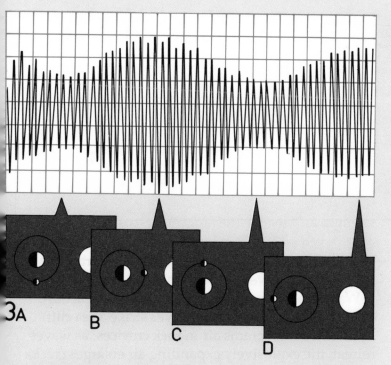

3 Tides (left)
A month's tides show how changes in tidal range follow changes in Earth, Moon, Sun alignment. Spring tides (high high tides and low low tides) and neap tides (low highs and high lows) each occur about every fortnight.
A Neap tide – Sun, Earth, Moon form a right angle.
B Spring tide – Sun, Earth, Moon form a straight line.
C Neap tide – Sun, Earth, Moon form a right angle again.
D Spring tide – Sun, Moon, Earth form a straight line.

3A   B   C   D

2 Underwater currents called undertows flow away from the shore, balancing the onshore pile-up of water by waves. Rip-currents are strong local currents of this kind.

3 Tides are two sets of huge progressive waves that sweep around the oceans each day. They are caused by the gravitational pull of the Moon and Sun. The Earth's spin, continents, coastlines and underwater ridges affect local tidal height. Coinciding with storm waves, the highest tides affect the highest level of the shore.

2 Underwater currents (below)
A Wave advances inshore against backwash from previous wave.
B Advancing wave peaks.
C Advancing wave breaks.
D Breaking wave thrusts water up the beach.
E Swash (water surging up the beach after a wave breaks)
F Backwash returns to sea, creating an undertow.

D   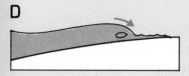   E   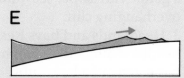   F

©DIAGRAM

141

Sea attacks the land

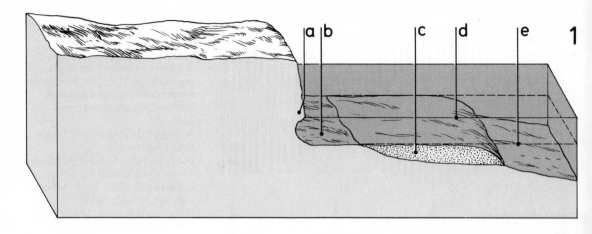

1

1 Cliff and wave-cut platform
a Sea cliff being formed by undercutting wave erosion
b Wave-cut platform
c Beach deposits
d High tide level
e Low tide level

2 Cliff steepness
A Gentle cliff slope influenced by landward-tilted rock layers
B Steep cliff influenced by seaward-tilted rock layers

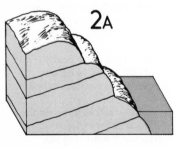

2A

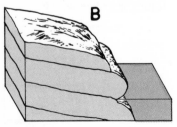

B

Where cliffs or rocks rim land the coast is probably retreating, as waves erode the shore.

Wave erosion works chiefly by hydraulic action, corrasion, and attrition. As waves strike a sea cliff, hydraulic action crams air in rock crevices; as waves retreat, the explosively expanding air enlarges cracks and breaks off chunks of rock. Chunks hurled by waves against the cliff break off more pieces – a process called corrasion. Rubbing against each other and the cliff reduces broken rocks to pebbles and sand grains – a process called attrition.

Different combinations of wave action, rock type, and rock beds produce these features.

1 Cliff and wave-cut platform A sea cliff forms where waves undercut a slope until its unsupported top collapses. As waves eat farther back inland they leave a wave-cut bench or platform jutting out beyond the cliff, below the sea.

2 Cliff steepness varies with rock hardness (usually the harder the rock the steeper the cliff) and the angle of rock layers. Landward-tilted layers tend to produce a gentle cliff slope; seaward-tilted layers may give an overhanging cliff.

3 Headlands and bays Resistant rock juts out as a headland after erosion of nearby less resistant rock has eaten out a bay. Subsequent erosion produces

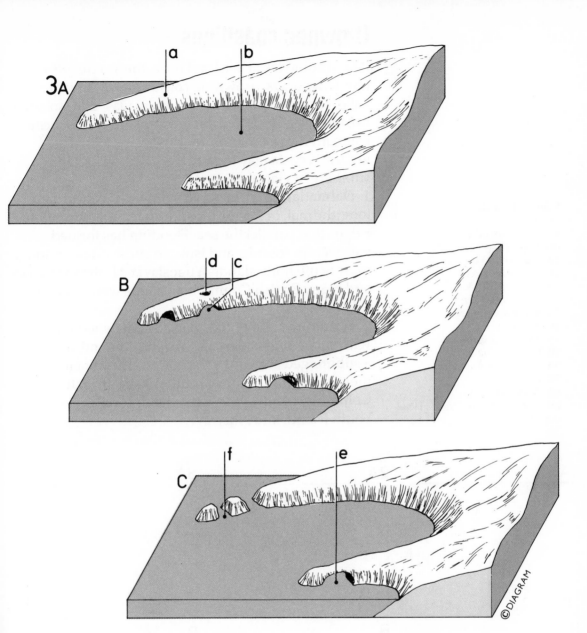

caves, blowholes, arches and stacks. At a sea cliff base, wave action may enlarge a horizontal crack to gnaw out a sea cave. Inside, erosion of a vertical joint may form a cliff-top blowhole. A sea cave driven through a headland forms an arch. Roof collapse isolates the headland's tip as a steep-sided island called a stack.

3 Headlands and bays
Diagrams show three stages in cliff erosion.
A Formation of headland (**a**) and bay (**b**)
B Erosion cuts sea caves (**c**) and cliff-top blowhole (**d**)
C Cave erosion creates arches (**e**) and, later, stacks (**f**)

Drowned coastlines

If land subsides or sea level rises, the sea invades low-lying areas, drowning coastal plains, invading valleys, converting ridges to peninsulas, and isolating uplands as islands. Submerged upland coasts include Dalmatian coastlines, fjords, and rias. Submerged lowland coasts may feature estuaries, fjards, and submerged glacial deposits.

1 Dalmatian coastlines (alias drowned concordant, longitudinal, or Pacific coastlines) feature mountain ridges that parallel the sea. Flooding has formed valleys into sounds, and isolated some ridges as long, narrow offshore islands. Yugoslavia's Dalmatian coast and many British Columbian islands were shaped this way.

2 Fjords are submerged, glacially deepened inlets with sheer, high sides, a U-shaped cross profile, and a submerged seaward sill largely formed of end moraine. Fjords occur in south Alaska, British Columbia, south Chile, Greenland, New Zealand's South Island, and Norway.

Drowned coastlines
Maps (**A**) and block diagrams (**B**) show corresponding types of drowned coastline.
1A Dalmatian Yugoslavia
1B Dalmatian coast
2A Southern Alaska
2B Fjord coast
3A Southwest Ireland
3B Ria coast
4A Thames rivermouth
4B Estuary coast
5A Southern Sweden
5B Fjard coast
6A Boston Harbor
6B Coast with submerged glacial deposits

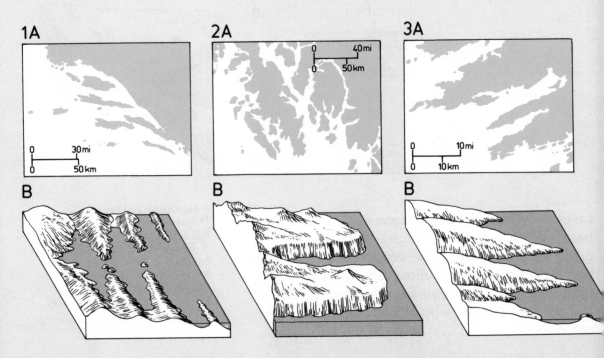

1A

2A

3A

B

B

B

3 Rias are drowned river valleys forming long, funnel-shaped, branching inlets meeting the sea at right angles and with a V-shaped cross profile. Rias abound in southwest England, southwest Ireland, northwest France, and northwest Spain.

4 Estuaries are tidal river mouths, many of them drowned low-lying river valleys, flanked by mud-flats and pierced by a maze of creeks submerged at high tide. Such estuaries include those of the Elbe, Gironde, Hudson, St. Lawrence, Susquehanna (Chesapeake), and Thames.

5 Fjards are ice-deepened lowland inlets often with small islands at the seaward end. They indent rocky, glaciated lowlands, as in southern Sweden, Nova Scotia, and the Shetland Islands.

6 Submergence of glacial deposits has left drumlins (see pp. 158–159) as offshore islands in Boston Harbor and Northern Ireland's Strangford Lough.

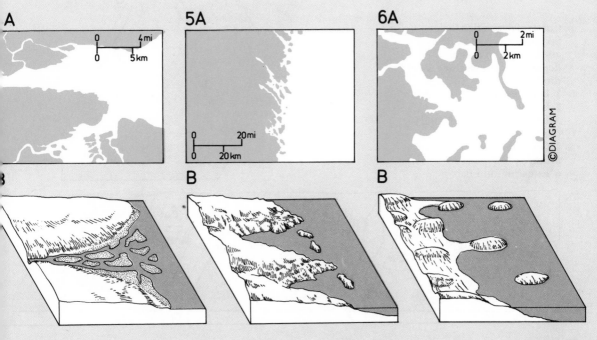

A

5A

6A

©DIAGRAM

B

B

How sea builds land

Rock fragments torn from one stretch of coast are often added to another. Storm waves may hurl seabed sand and shingle inshore. Waves breaking at an angle to the land produce the longshore (littoral) drift of sand and shingle along the beach. Where land slopes gently to the waves extensive beaches form, and land may grow into the sea. But storms shift or strip away huge quantities of loose material, especially on steeply sloping upper beaches. In the end, gravity transfers most beach material to the sea bed.

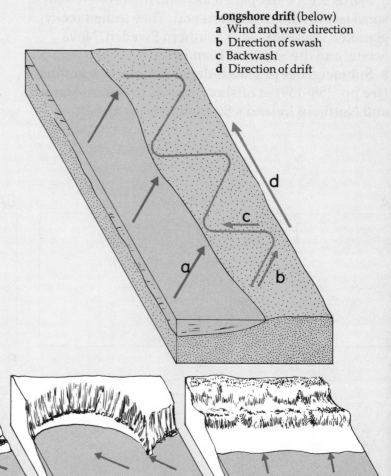

Shore deposition (above)
On a gently sloping shore (**a-b**) sea erodes the lower part (**c**) and dumps it on the upper part (**d**). On a steeply sloping shore the opposite occurs. Most beaches are in near-equilibrium and any change occurs very slowly.

Longshore drift (below)
a Wind and wave direction
b Direction of swash
c Backwash
d Direction of drift

Coasts of deposition (below)
Six illustrations show items associated with coastal deposition. Arrows show wave direction.
1 Boulder beach
2 Bay-head beach
3 Lowland beach, with sand dunes blown inshore
4 Offshore bars
5 Spit
6 Estuarine mud-flats

1

2

3

Coasts of deposition include these items:

1 Boulder beaches are narrow belts of rocks and shingle at the base of sea cliffs.

2 Bay-head beaches are small sandy crescents. Each lies in a cove between two rocky headlands.

3 Lowland beaches are broad, gently sloping sandy beaches with a strip of shingle on the upper shore, often backed by dunes of sand blown inland by onshore winds. Such shores rim the southern Baltic and form the Landes of southwest France.

4 Bar A bar is an offshore strip of sand or shingle parallel to the coast. Bars border most of the US southeast and Gulf coastline. They include the barrier islands that form Cape Hatteras.

5 Spit A spit resembles a bar, with one end tethered to the land. Special forms include bay-bars that link two headlands, tombolos linking islands to a mainland as in England's Chesil Beach, and (some) cuspate forelands: triangular shingle formations such as Cape Canaveral and Dungeness, England.

6 Mud-flats and **salt-marshes** form in estuaries (see pp. 144–145) and sheltered bays. Mud-flats are barren beds of silt and clay deposited by tides and drowned at high tide. Silt-trapping plants raise their level, creating salt-marsh drained by tidal creeks. In the tropics, mud-flats colonized by mangrove trees form mangrove swamps.

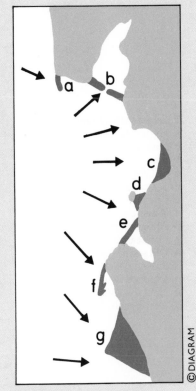

©DIAGRAM

Depositional features (above)
a Spit
b Double spits
c Bay-head beach
d Tombolo
e Barrier beach
f Hooked spit
g Cuspate foreland

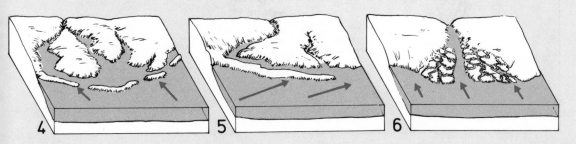

Shores risen from the sea

While sinking land or rising sea level has drowned some coasts, rising land or falling sea level has stranded other coasts above the present level of the waves. Raised highland and lowland coasts have somewhat different characters.

1 Emergent highland coasts feature raised beaches. A raised beach, often covered with shells or shingle, stands perched high and dry above sea level. The raised beach is an old shoreline and adjoining wave-cut rock platform up to 2600ft (about 800m) across. Inland, above the raised beach, rises an old sea cliff, perhaps pierced by wave-cut caves. Below the raised beach, a new sea cliff and wave-cut platform form the seaward boundary. Successive earth movements can create a series of raised beaches, one above the other. Many formed where land depressed by ice bobbed up once the ice had melted. Such

1 Emergent highland coast
a High tide level
b Low tide level
A Before emergence
c Sea cliff
d Sea caves
e Beach
f Wave-cut platform
B After emergence
g Old sea cliff
h Old sea caves
i Raised beach
j Exposed wave-cut platform
k New sea cliffs
l New beach

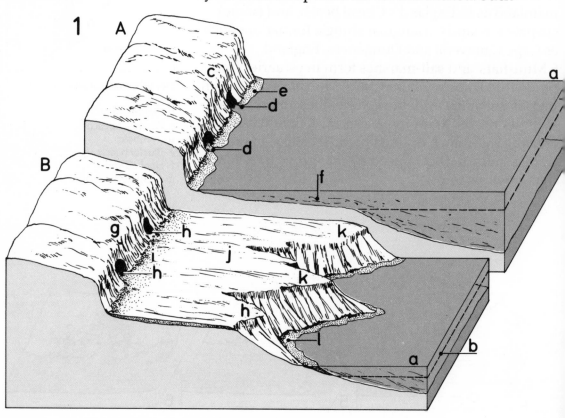

beaches occur on North America's Arctic coasts, in western Scotland, and around the Baltic Sea. Others crop up as far apart as Malta and the South Pacific.

2 Emergent lowland coasts are coastal plains sloping gently to the sea, here rimmed by marshes, sandy beaches, bars, lagoons, and spits. The plains are uplifted continuations of the shallow offshore continental shelf, and thus may be floored by seashells, sand, and clay consolidated into limestone, sandstone, and shale. Inland the plains may end abruptly below a line of hills marking the old coastline. This happens in the southeast United States: its coastal plain abuts the Fall-line where rejuvenated rivers flow steeply from the Appalachian Mountains to create a line of waterfalls. Other emergent lowland coasts include the northern Gulf of Mexico, the southern Rio de La Plata, and much of east-coast India.

2 Emergent lowland coast
A Before emergence
a Gentle slopes
b River valleys
c Coastline
d Continental shelf
B After emergence
e Valleys deepened by rejuvenated rivers
f Old coastline
g New coastal plain
h New coastline

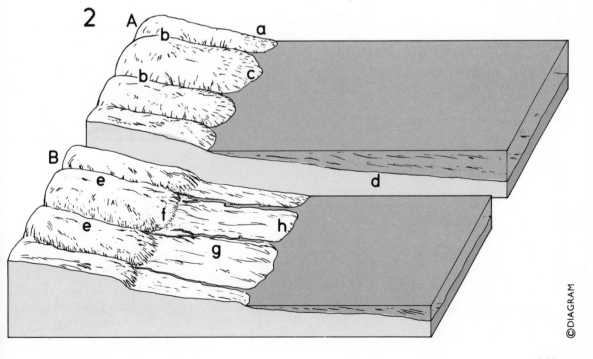

Where coral grows

Coral reefs rim shores and form low islands mainly in the world's warm seas and oceans. Coral is a limestone rock produced by tube-shaped skeletons of billions of coral polyps – animals resembling tiny sea anemones – and by limy algal plants called nullipores. Reefs grow up and out as new organisms build on the skeletons of old. Waves pounding the seaward edge hurl broken chunks of coral on the reefs. Meanwhile sand accumulates upon the shoreward side. So reefs become low islands.

Coral polyps need clear, warm, shallow, salty water, found mostly in tropical seas and oceans to the east of continents. But many prehistoric reefs now lie inland in rocks outside the tropics.

There are three main types of coral reef.

1 Fringing reefs are narrow offshore coral reefs, separated by a shallow lagoon from the nearby coast. Such reefs abound off the Bahamas and Caribbean islands.

1 Fringing reef
a Edge of land
b Narrow, shallow lagoon
c Living coral
d Mound of dead coral
e High water
f Low water

2 Barrier reef
a Edge of land
b Wide, deep lagoon
c Living coral
d Mound of dead coral
e High water
f Low water

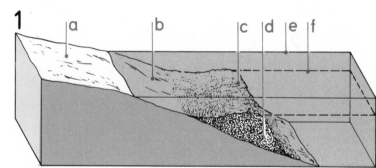

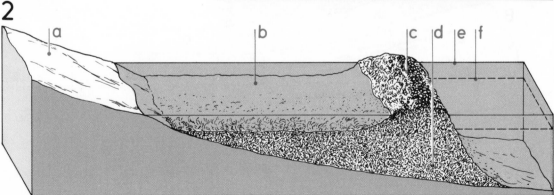

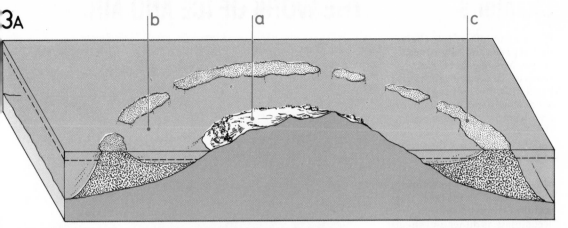

3A

b a c

2 Barrier reefs are broad coral platforms separated from the coast by a wide, deep channel. At least some formed upon subsiding coasts. The world's largest is northeast Australia's Great Barrier Reef, 1260mi (2027km) long.

3 Atolls comprise circular chains of coral reefs. Each atoll encloses a lagoon and probably started as a fringing reef around a volcanic island. As the island began to sink and/or sea level rose, coral growth kept pace. So the reef became a barrier reef and then an atoll. Borings show that Eniwetok Atoll in the Pacific Ocean grew upward from a volcanic base now more than 4600ft (1400m) below the surface. Atolls abound chiefly in parts of the Pacific and Indian oceans. The Pacific's Marshall Islands include the world's largest atoll, Kwajalein, more than 170mi (274km) long.

3A Atoll forming
a Slowly sinking volcanic island
b Lagoon
c Barrier reef

3B Atoll formed
a Submerged volcanic island
b Lagoon
c Coral reef

3B

b a c

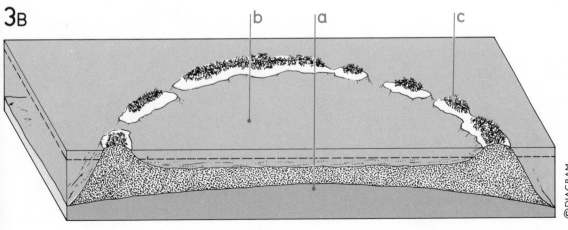

©DIAGRAM

Chapter 9   THE WORK OF ICE AND AIR

Ice frets mountains into sharp-tipped peaks and saw-edged ridges. Moving sheets and rivers of ice deepen valleys and bevel hills. But melting ice sheds sheets and strips of debris on lowlands. Around an ice sheet's rim, alternate freezing and thawing rework the surface of the ground.

Where wind scours deserts, blown sand sculpts rocks and wears hollows in the land. But windblown particles accumulate as dunes and sheets. The work of wind and water creates distinctive desert landscapes.

Three panoramic views of the Swiss Alps. (Engravings taken from an old edition of *Baedeker's Guide to Switzerland*)

153

Glaciers and ice sheets

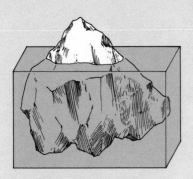

Castellated berg (above)
Such bergs plunge into the sea
from Arctic glaciers. Vast flat-
topped bergs break off the ice
shelves that fringe Antarctica.
Polar seas also bear vast tracts of
sea ice.

Ice covers about 10 per cent of all land and 12 per cent
of the oceans. Most lies in polar sea ice, polar ice
sheets and ice caps, valley glaciers, and piedmont
glaciers formed by valley glaciers merging on a plain.

Most of Antarctica lies beneath an ice sheet twice
the size of Australia and up to 14,000ft (4300m) thick.
Another ice sheet covers Greenland. Such vast slabs
of ice spread slowly, smothering all but a few
projecting peaks, or nunataks. Smaller ice masses,
called ice caps, crown parts of Iceland, Norway, and
some Arctic islands.

Valley glaciers are tongues of ice in mountain
ranges. They start in ice-worn rock basins called
cirques (see also pp. 156-157). Here, old snow forms
firn or nevé – a mass of ice pellets compacted by the
weight of snow above. Fed by fresh snowfalls, firn

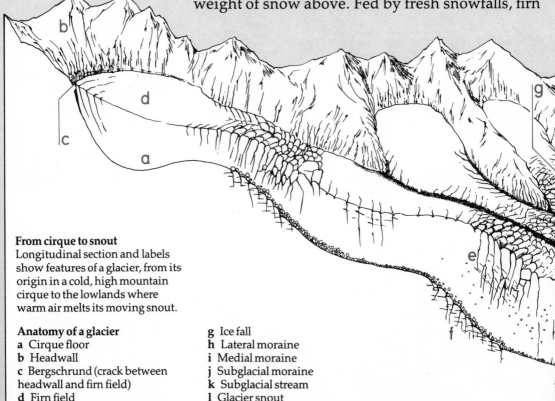

From cirque to snout
Longitudinal section and labels
show features of a glacier, from its
origin in a cold, high mountain
cirque to the lowlands where
warm air melts its moving snout.

Anatomy of a glacier
a Cirque floor
b Headwall
c Bergschrund (crack between
   headwall and firn field)
d Firn field
e Crevasses
f Rock step

g Ice fall
h Lateral moraine
i Medial moraine
j Subglacial moraine
k Subglacial stream
l Glacier snout
m Terminal moraine

spawns a glacier that spills down a valley filling it with ice, perhaps for scores of miles.

Glaciers creep downhill at an inch (2.5cm) to 100ft (30m) a day, according to conditions. Pressure makes the lower ice plastic enough to flow, but the upper ice stays rigid. Varying gradients and differential rates of flow within the glacier split the surface with deep cracks called crevasses. Where a glacier plunges down a steep rock step the surface forms an ice fall – a criss-cross maze of crevasses and isolated pinnacles.

Frost-shattered rocks and stones falling on the glacier's flanks form lines of lateral moraines. Where a tributary glacier joins a major glacier two lateral moraines merge as a medial moraine.

Most valley glaciers end in a moraine-rich snout where warm air melts ice as fast as the glacier flows. If ice melts any faster, the snout retreats.

In polar regions, glaciers and ice sheets extend down to the sea. Huge chunks of ice snap off and float away as icebergs. But Antarctica's Ross Barrier is a floating slab of ice as big as Spain, tethered to the land.

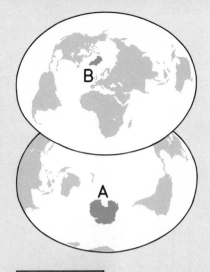

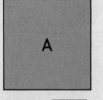

Frozen landmasses (above)
Maps locate both landmasses mostly under ice.
A Antarctica
B Greenland
Block diagrams compare their ice sheets with the areas of two countries.
C United States
D Mexico

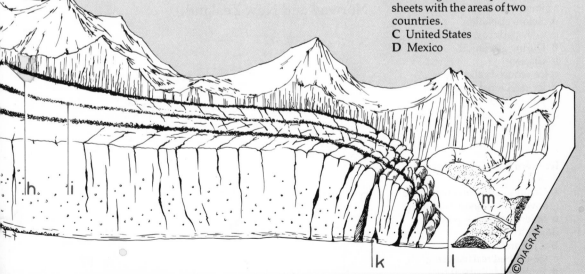

How ice attacks the land

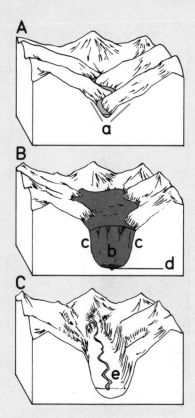

When glaciers and ice-sheets melt they leave a landscape rasped by moving ice. Ice sheets armed with broken rocks bevelled huge tracts of northern Canada and Finland. Frost and moving ice have sharpened peaks and deepened mountain valleys in the Andes, Alps, Himalayas, and Rockies.

Valley-glacier erosion starts high up on hills or mountainsides. Here, freeze-thaw processes fracture rocks, enlarging and deepening shallow snow-filled dips to form the steep rock basins known as corries, cwms, or cirques. Two cirques eating back into a mountain may leave a knife-edged ridge called an arête, sometimes notched to give a pass or col. Where at least three cirques converge, arêtes meet in a pyramidal peak. The famous Matterhorn originated in this way.

A glacier that fills a mountain valley shoves loose material ahead, and plucks rocks from the valley sides. Loosened stones embedded in the frozen river's flanks rasp more rock from the valley sides. Thus a glacier widens, deepens, and straightens – changing a valley's V-shaped cross section into the U-shape of the glacial trough. (Glacial troughs invaded by the sea became the fjords of Alaska, Norway and New Zealand.)

Gnawed by ice (above)
Three diagrams picture major changes wrought by glaciation in a mountain valley.
A Before glaciation
a V-shaped cross section
B During glaciation
b Glacier
c Ice-scoured valley sides
d Ice-deepened valley floor
C After glaciation
e Glacial trough with U-shaped cross section

Ice-worn hummocks (right)
A *Roche moutonnée*
a Gently sloping upstream side, grooved by stones in moving ice
b Steep, rough, ice-plucked downstream side
B Crag and tail
c Ice-rubbed resistant crag
d Protected tail of soft rock

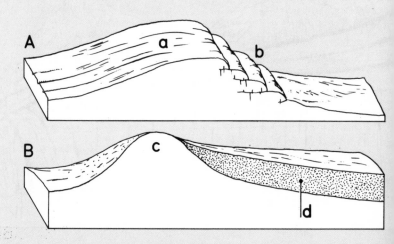

Glacial erosion truncates (lops off) spurs that had jutted out into the main valley and leaves tributary streams plunging from the lips of hanging valleys ending high above the bottom of the trough. Yosemite's Bridalveil Fall is such a waterfall.

Rock debris embedded in moving ice scratches and polishes the valley walls and floor, producing tell-tale grooves. (Similar striations betray the ice sheet that once covered New York's Central Park.) Other tell-tale features include two distinctive types of ice-worn hummocks known as crag-and-tail and *roche moutonnée*.

Glaciated land (below)
Layers pinpoint seven features seen in glaciated mountains and valleys.
a Col
b Cirques
c Arêtes
d Pyramidal peak
e Glacial trough
f Truncated spurs
g Hanging valley

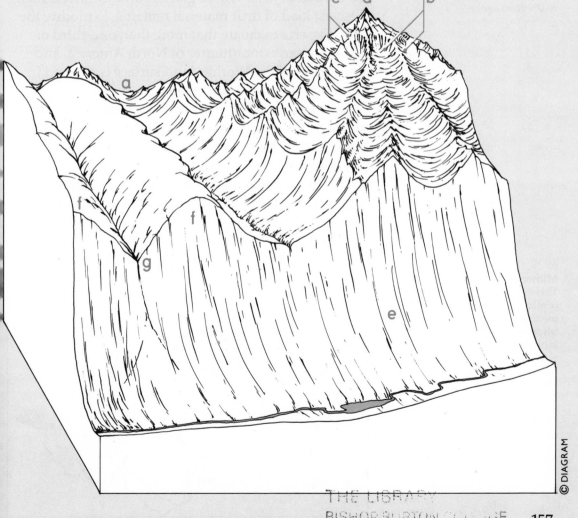

© DIAGRAM

Erratic block (above)
An immense granite boulder taller than a man lies on limestone. A melting ice sheet dumped the boulder more than 10,000 years ago.

On, in, and beneath their ice, glaciers and ice sheets shift substances from finely powdered rock to mighty boulders. Streams emitted by the ice transport more ice-eroded debris. Where glaciers and ice sheets melt, this vast load of drift material remains, to modify the land. Experts calculate that more than one-third of Europe, nearly one-quarter of North America, and one-eighth of the World's land surface is cloaked in debris shed by ice or meltwater.

Debris dumped by glaciers themselves is an unsorted mass of stones and rock embedded in a sandy, clayey matrix and known as till or boulder clay. Some till forms under active ice, some accumulates where ice decays. How and where till forms determines landscape features in huge tracts of lowland. Retreating ice left the undulating sheets called ground moraine that cover much of the North European Plain. Drift added flat floors to many

Midwest drift deposits (right)
This map marks southern limits of glacial deposits laid down by melting ice sheets in the US Midwest. Each line marks the farthest advance of a different glacial stage, from Nebraskan (early) to Wisconsin (late).

- —— Wisconsin
- —— Illinoian
- —-–- Kansan
- —-–- Nebraskan

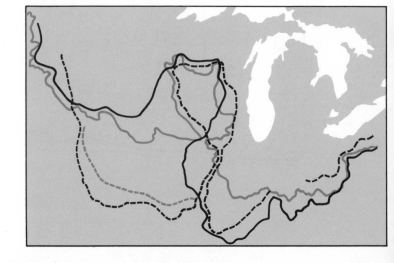

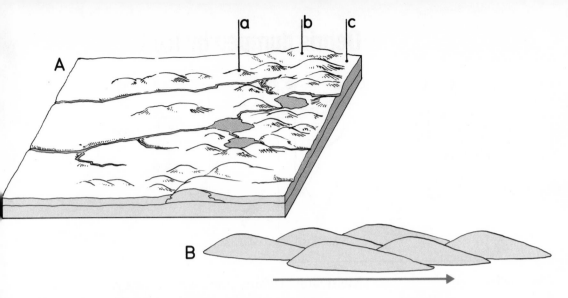

U-shaped glaciated valleys. And where an ice front paused you find the long, curved ridges of its terminal or end moraines.

Drumlins are elongated hummocks up to 300ft (90m) high and 1mi (1.6km) long. Many formed where valley glaciers shed and streamlined drift as they reached a plain and spread out.

Erratic blocks had other origins. Many of these ice-borne rocks lie far from where they started. Some Scottish rocks have ended up in southeast Ireland, and 600mi (1000km) separate Kentucky's red jasper boulders from their nearest possible source north of Lake Huron. Erratic blocks can be enormous. One Albertan specimen reportedly exceeds 18,000 tons.

Glacial deposits
A The block diagram (above) shows a landscape largely formed by drift
a Drumlins
b Terminal moraine
c Outwash sands
B A swarm of drumlins aligned with ice flow (arrow)

North European moraines
Here we show southern limits of glacial deposits laid down by successive, melting ice sheets based on Scandinavia. Terminal moraines formed where a melting ice front paused.

~~~. Recent terminal moraines
~~~ Earlier terminal moraines
••••••• Southern edge of glaciation

Debris dumped by ice 2

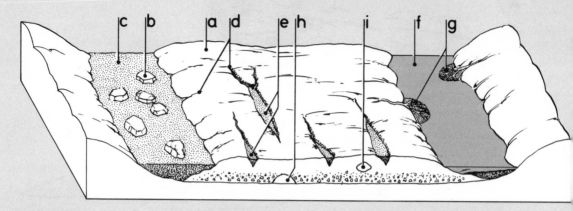

Glacial valley (above)
a Glacier
b Blocks of melting ice
c Lake filled by sediment
d Surface streams
e Crevasses containing sediment
shed by streams
f Ice-margin lake
g Deltas
h Subglacial stream
i Englacial stream

Postglacial valley (below)
a Exposed valley floor
b Kettles (ex ice-filled hollows)
c Kame terrace (old lake bed)
d Streams
e Kames (ex crevasse deposits)
f Kame terrace (old lake bed)
g Kame-deltas
h Esker from subglacial stream
i Esker from englacial stream

Meltwater streams that issue from a glacier or ice sheet produce layered sediments. These therefore differ from the unsorted drift directly dropped by ice. Geologists call them outwash or glaciofluvial deposits.

Outwash comes in two main forms: outwash plains and valley trains. Outwash plains are layered sheets of clay, sands, and gravels that fan out over lowlands ahead of where an ice-sheet lay. The largest particles settle near the ice rim. Finer particles may travel many miles. Valley trains are thick outwash deposits covering the floors of deep, narrow, glaciated valleys.

Outwash deposits give rise to the following features:

1 Eskers These long, winding ridges are aligned with the flow of retreating glaciers or ice sheets. They are from a few feet to a few hundred feet wide, up to

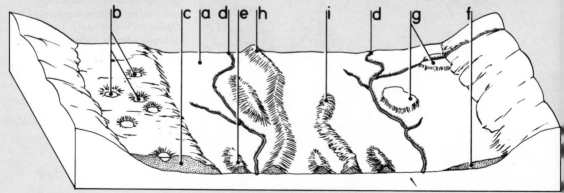

100ft (30m) high and some extend for many miles. They grew where subglacial streams shed loads in ice tunnels or at their receding mouths. Eskers are plentiful in lowland areas around the Baltic Sea.

2 Kames are mounds that come in two main forms called kame-terraces and kame-moraines. A kame-terrace is a narrow, flat-topped, steep-sided ridge of sediments along a valley side. It formed below a stream or lake trapped between the valley side and a prehistoric glacier. Kame-moraines or kame-deltas are complex undulating mounds of sands and gravels dumped along a stagnant ice sheet's rim. Kame-moraines are plentiful in central Ireland and the United States between Long Island and Wisconsin.

3 Kettles are hollows in kame-terraces or kame-moraines, formed where ice blocks melted. Such ice blocks occur today in Iceland.

4 Varve clays are banded layers of fine and coarse material deposited in meltwater lakes fringing ice sheets. Coarse material was washed in with the summer thaw. Fine material settled in winter when the lakes froze over. Counting bands enabled geologists to measure post-glacial time year by year in North America and Sweden.

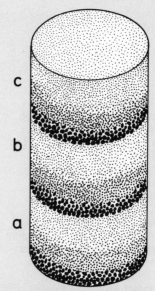

Varve clays (above)
This core sample shows three banded layers, representing sediments laid down in three successive years in a lake liable to winter freezing.
a First year
b Second year
c Third year

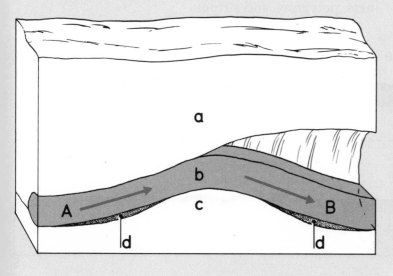

Sub-glacial sediments (left)
Here we show two situations where a stream beneath a glacier may shed its load.
A The stream flows uphill and loses carrying power.
B The channel widens, so water flows at reduced pressure.
a Ice
b Stream
c Bedrock
d Sediment

Around an ice sheet's rim

Prolonged freezing and brief summer thawing around an ice-sheet's rim produces the periglacial ("around the ice") or tundra landscapes of far northern Eurasia and North America. Some periglacial features dating from the last glaciation still show in lands much farther south.

A major influence is permafrost: permanently frozen ground beneath the surface. In places, Siberian permafrost extends 2000ft (610m) deep. Above it lies the active layer, 6–20ft (2–6m) thick. In summer, meltwater fills and lubricates the active layer, so sloping surfaces creep downhill upon the frozen layer beneath. Known as solifluction, this process dumps so-called head deposits of frost-shattered rocks, like the chalk-rubble coombe deposits still seen in parts of southern England.

Freeze-thaw affects both solid rock and loose materials. On north-facing slopes, freeze-thaw beneath snow patches loosens bits of bedrock, and meltwater and solifluction carry them away. Called nivation, this sequence wears big nivation hollows in the rock. Below loose surfaces, freezing water expands, so the ground heaves. Repeated heaving sorts out large and small soil particles. The result is patterned ground with stones arranged in circles, nets, polygons, and stripes.

Permafrost (below)
Map and diagrams contrast the relative extent of permafrost in parts of Asia and North America.
a North Pole
b Arctic Circle
c Northeast Asia
d North America
A Cross section of Alaska showing north-south extent and depth of permafrost (**e**) and active layer (**f**). (Length and depth are drawn to different scales.)
B Similar cross section of northeast Asia

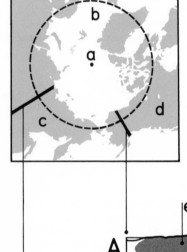

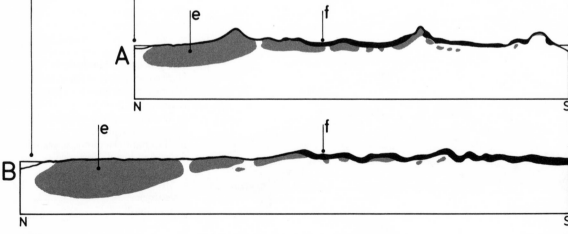

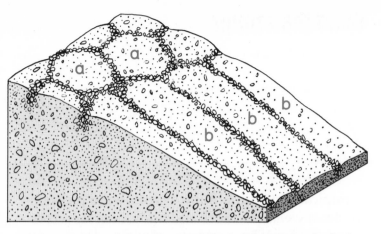

Patterned ground (left)
a Stone polygons one yard (1m) or more across, with sorted fine material inside
b Stone stripes: parallel lines of stones formed on steep slopes under the influence of soil creep

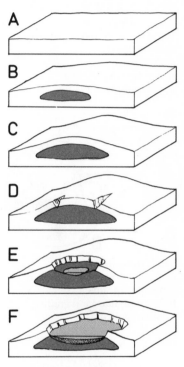

Other hallmarks of the periglacial fringe include ice wedges, thermokarst, and pingos.

Ice wedges form where cracks appear in frozen ground. Meltwater fills the cracks and freezes as ice wedges that taper downward for as much as 35ft (10m). Gravel often fills old ice wedges.

Where ground ice melts it often leaves a surface pocked by hollows, superficially like a karst limestone surface, and thus called thermokarst.

The strangest periglacial features are pingos: cones or domes 20–300ft (6–90m) high. These earth or gravel humps occur where freezing water expands between an almost frozen surface and the permafrost, pushing up an earth or gravel blister.

Ice wedge (left)
Erosion of a subarctic river bank reveals a V-shaped ice wedge taller than a man, formed by water freezing in a crack in layered silt.

Pingo formation (above)
A Pre-pingo land surface
B Ice lens forms underground.
C Expanding ice lens pushes up a dome in the land surface.
D Tension cracks in the dome expose the ice lens.
E Melting ice creates a pond.
F Sediment collects on the pond floor.

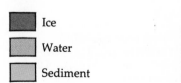

■ Ice
▨ Water
□ Sediment

© DIAGRAM

163

Wind the eroder

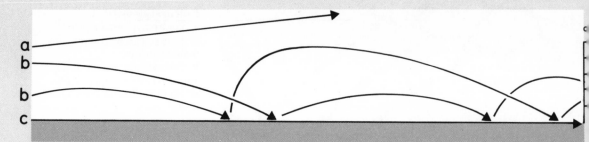

Wind the transporter (above)
Wind moves different particles at different levels.
a Dust particles
b Sand grains
c Tiny pebbles

In tropical and mid-latitude deserts, wind picks up specks of weathered rock and hurls them far across the barren land. Fine particles of silts and clays whirl high above the ground in dust storms. Sand grains hop and skip across the countryside. Tiny pebbles roll and slide. Of all these processes, sandblasting does most to bevel desert rocks. Attacking pebbles, soft rocks, and rock crevices, windblown desert sand creates the following erosion features.
1 Ventifacts are stones with surfaces smoothed and flattened under prolonged attack by windblown sand.
2 Rock pedestals are mushroom-shaped rocks, often formed of horizontal rock layers. Sand gnaws into their bases, but winds are seldom strong enough to lift sand grains above waist height. (Sand can similarly cut rock caves.)

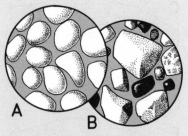

Sands compared (above)
A Polished, rounded, "frosted" desert sand grains
B More angular sand grains, from a river bed

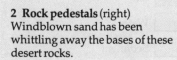

1 Ventifacts (above)
Sand-blasted pebbles with flat, smooth facets (also called *dreikanter*)

2 Rock pedestals (right)
Windblown sand has been whittling away the bases of these desert rocks.

164

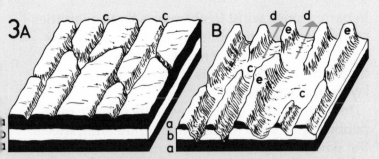

Windblown deposits

3 **Zeugen** (left)
A Rock beds before erosion
B Rock beds after erosion
a Hard rock
b Soft rock
c Joints
d Wind direction
e Zeugen

3 **Zeugen** are parallel, flat-topped ridges of hard rock up to 100ft (30m) high. They are left standing when sand has widened joints in horizontal hard rock and gnawed into the softer rock beneath.

4 **Yardangs** are parallel ridges of hard rock up to 50ft (15m) high. They form where alternating hard and soft rock layers were upended. Wind gnaws the soft rock into furrows, but leaves the hard rock standing. Yardangs occur especially in Central Asia and northern Chile's Atacama Desert.

5 **Rock pavement** or **hamada** is a flat, wind-smoothed rocky desert surface (not shown).

6 **Deflation hollows** are worn or deepened in a desert surface by the wind. Egypt's Qattara Depression – the world's largest deflation hollow – is about as big as Wales, and as much as 400ft (121m) below the level of the sea. It is partly a tectonically formed feature, further eroded by the wind. Southwest Africa, Western Australia, and Mongolia have smaller wind-worn "saucers." Some expose supplies of underground freshwater, so producing swamps, or lakes, or supporting fertile oases.

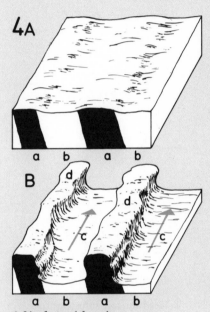

4 **Yardangs** (above)
A Rock beds before erosion
B Rock beds after erosion
a Hard rock
b Soft rock
c Wind direction
d Yardangs

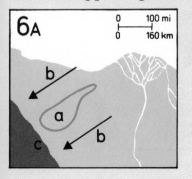

6 **Deflation hollows**
A Left: Map of Northern Egypt
a Fayum Depression
b Prevailing winds
c Sand removed by wind
B Right: Diagram of hollow
a Deflation hollow
b Faulted rock layers (faulting predates erosion)
c Wind direction

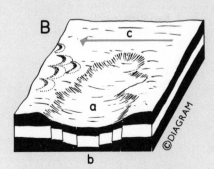

165

Windblown deposits

In barren lands wind freely moves huge quantities of tiny particles eroded from the rocks. In deserts and near prehistoric ice sheets these windblown fragments cloak vast areas with shifting sands or layers of cemented dust.

In some deserts winds slowed by passing over pebbles shed sand in smooth or undulating sheets. But winds blowing steadily from one direction pile sand into the mobile hillocks known as dunes. These three are the best-known forms.

1 Head and tail dunes grow in dead air spaces near a rock or shrub. The long leeward tail can grow to almost half a mile (750m) long.

2 Barkhans are dunes with low, curved flanks like horns, blown forward faster than the middle. Some barkhans grow 100ft (30m) high. Each migrates as wind pushes sand across its crest.

3 Seif ("sword") dunes form long, wavy, ridges up to 700ft (215m) high, thrown up by vortices in a prevailing wind, or where barkhans are elongated by a cross wind.

1 Head and tail dune (right)
a Obstacle impeding flow of windblown sand
b Head dune on windward side of obstacle
c Tail dune on leeward side
d Wind direction

2 Barkhans (right)
a Gentle windward face
b Steep leeward face
c Horns
d Prevailing wind
e Eddy

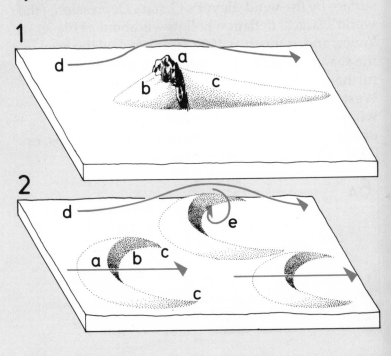

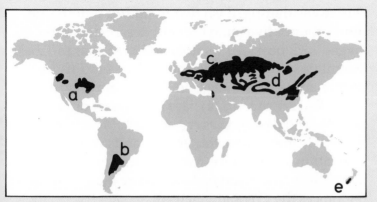

Loess deposits (left)
World map locates loess and
loess-like deposits in four
continents and New Zealand.
a North America
b South America
c Europe
d Asia
e New Zealand

Windblown particles finer than sand settle far
beyond their point of origin. Loosely cemented
silt-sized grains from Mongolia's Gobi Desert form
layers up to 1000ft (300m) thick in northern China.
Geologists call these deposits loess from a German
word for "loose". Known as adobe in the US, and as
limon in France, similar material covers parts of
North America and Europe where winds blew dust
from sands and clays deposited by ice sheets in the
Pleistocene.

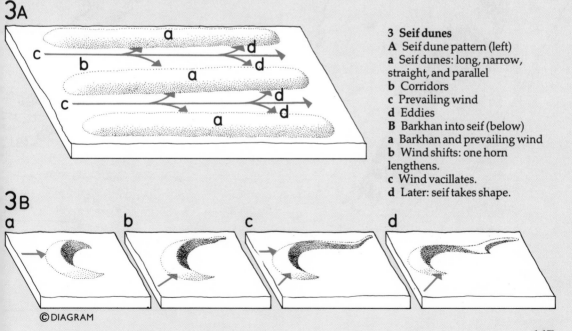

3A

3 Seif dunes
A Seif dune pattern (left)
a Seif dunes: long, narrow,
straight, and parallel
b Corridors
c Prevailing wind
d Eddies
B Barkhan into seif (below)
a Barkhan and prevailing wind
b Wind shifts: one horn
lengthens.
c Wind vacillates.
d Later: seif takes shape.

3B

©DIAGRAM

Lands shaped by wind and water

Desert weathering, flash floods, and/or windborne sand produce five main types of desert landscape: (**1**) Jagged rock peaks, as in Sinai and the Sahara's Tibesti Mountains; (**2**) Desert plateau with steep cliffs and deep, narrow river valleys; (**3**) Stony, gravelly, desert, also called *reg* or *serir*; (**4**) bare rock desert or *hamada*; and (**5**) sand desert, alias *erg* or *koum*. Most have harsher features than you find in lands where vegetation softens the effects of sun, wind, frost, or rain. These five desert features figure mainly in such regions as the Colorado Plateau:

1 Mesa Flat-topped, steep-sided plateau of horizontal strata capped by erosion-resistant rock.

2 Butte Isolated flat-topped hill, like a mesa but smaller.

Eight desert features
Numbered items in this diagram correspond with items featured in the text.
1 Mesa
2 Butte
3 Inselberg
4 Pediment
5 Canyon
6 Wadi
7 Alluvial fan
8 Playa

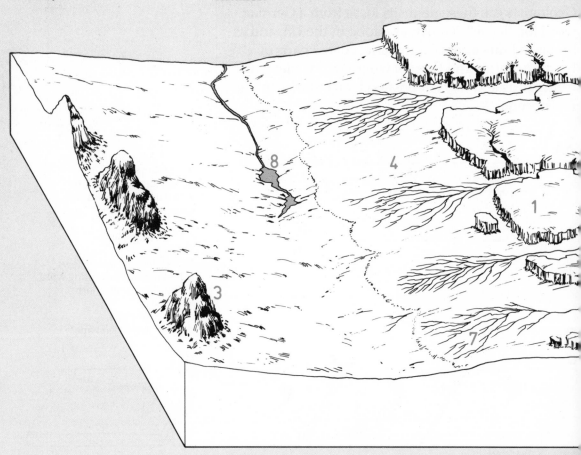

3 Inselberg Steep, isolated, hill with a narrow summit.

4 Pediment Gentle slope often covered with loose rock and lying below a mesa, butte, inselberg, or ridge. Pediments seem to be formed by weathering and floodwater.

5 Canyon Deep gorge of a river, often one flowing through a desert but fed by water from outside.

6 Wadi Usually dry desert watercourse, also called an *arroyo*, wash, or *nullah*.

7 Alluvial fan Fan-shaped mass of alluvial deposits shed by a fast-flowing mountain stream entering a plain or broad valley.

8 Playa (salt pan) Temporary brackish lake, such as many found in Nevada and Utah.

The world's deserts
Arid (**A**) and semiarid (**B**) areas lie mainly in the dry hearts of continents.
a North American Desert
b Peruvian-Atacama Desert
c Patagonian Desert
d Sahara Desert
e South-west African deserts
f Middle Eastern deserts
g Central Asian deserts
h Australian Desert

A
B

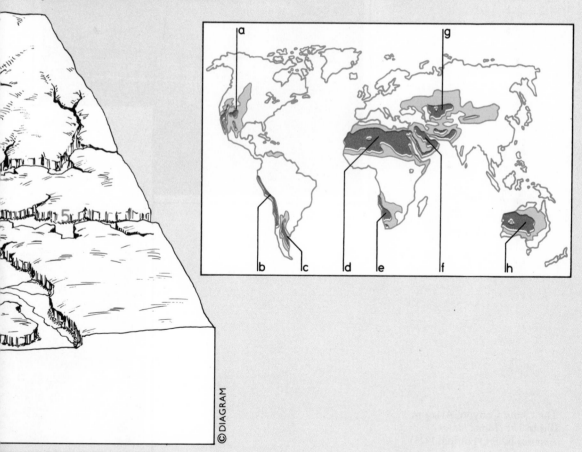

©DIAGRAM

Chapter 10

CHANGE THROUGH THE AGES

Processes producing and destroying land have shaped the crust ever since our planet gained its solid rocky skin.

Geologist-detectives can now work out the broad sequence of events. Radiometric dating and relative dating based on layered rocks and fossils help scientists read the story in the rocks. Its first, longest, and least-known volumes were the Archean and Proterozoic eons.

The Grand Canyon, Arizona. (From *The United States of America* by E O Hoppé, 1928)

Relative dating: using rocks

Which formed when? (right)
Volcanic ash (**a**) fell on limestone (**b**) laid down on glacial debris (**c**) dumped on slate (**d**) changed from pre-existing shale by granite (**e**). A diabase (dolerite) dike (**f**) pierced **d** and **c** before **a** and **b** formed. Glacial erosion explains the unconformity at **g**.

Clues to age sequences (below)
Youngest rocks are at the top.
A Mud cracks
B Ripple marks
C Graded bedding
D Cross bedding
E Pillow lava

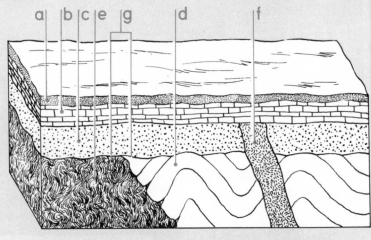

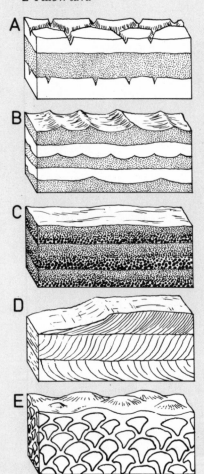

Earth's history lies locked up in the rocks that form its crust. Sedimentary rock layers or strata were laid down on top of one another, like pages in a history book. Reading these pages is the study called stratigraphy. Much of this is based on studying the properties of rocks themselves.

Geologists identify individual strata largely by such properties as grain size, minerals, and color. These features and distinctive fossils (see p. 176) help experts to define rock units, or formations. Geologists may divide formations into members and beds or lump them together in subgroups, groups, or supergroups. (For time rock units see p. 180.)

Many pages in Earth's "book" are torn, turned upside down, displaced, or lost. Stratigraphy involves working out the true sequence in which rock strata formed in any given place, matching these with layers elsewhere, and noting local gaps where erosion has wiped strata from the record.

Various clues help rock detectives discover where earth movements or injected molten rock have tampered with the evidence. For instance, steeply sloping strata never formed that way. Faults, folds, and injections of molten rock are younger than the rocks they affect. Mud cracks, ripple marks, and pillow lava create distinctive patterns on a layer's upper surface,

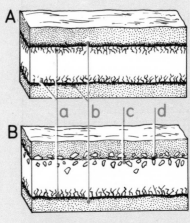

which becomes its base if the rock is overturned. A break between level rock layers above and crumpled layers below is an unconformity, hinting at a time gap when rock layers vanished by erosion.

Other clues help experts correlate the age of rocks formed at the same time in different places. Widely separated strata may share a unique set of characters, or facies. Migrating shorelines may mark a worldwide fall or rise in sea level. Widespread layers of volcanic ash could hint at an immense volcanic eruption. Lavas or sediments accumulating at the same time lock in particles aligned in the same direction by the Earth's magnetic field which has undergone a sequence of reversals. Matched alignments and matched fossils help geologists to correlate the relative ages of rock-cores sampled from around the world.

Migrating shoreline (above)
Facies change vertically here where rising sea level moved a shoreline to the right.
A Sea
B Beach
C Lagoon
a Limestone
b Fine sand
c Coarse sand
d Windblown sand
e Mud

Paleomagnetism (below)
As this volcano erupted, magnetic minerals in the lava aligned to match lines of force of Earth's magnetic field.
a Magnetic north
b Lava
c Lines of force
d Aligned minerals, enlarged

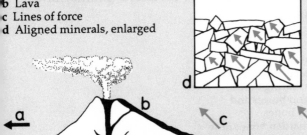

Sill or buried lava flow? (above)
The following clues help geologists tell a sill (**A**) from a buried lava flow (**B**).
a Chilled margin
b Baked contact
c Weathered lava surface
d Eroded bits of lava in overlying bed

©DIAGRAM

173

Relative dating: fossils 1

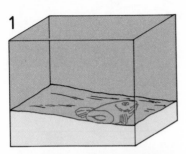

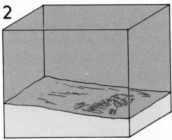

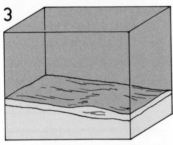

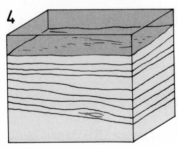

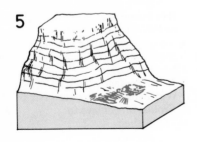

What we know of the relative ages of rocks owes much to paleontology – the study of fossils. Fossils are remains of prehistoric organisms, locked in sedimentary rocks being laid down when those organisms died.

Most dead organisms soon rot away. Fossils tend to be hard parts like wood or bone quickly buried and preserved by sediment below a sea or lake. Percolating minerals may reinforce a fossil. Or a corpse may dissolve to leave a hollow called a mold. Minerals that fill a mold create a cast. Plant leaves leave carbon films. Even burrows, tracks, and droppings can be fossilized.

In time Earth movements lift fossil beds above the sea. Rivers cut down through the topmost bed, exposing lower layers. So paleontologists find fossils formed in rocks laid down at different times.

These studies show that living things evolved through time. Minute one-celled organisms gave rise to many-celled plants and animals. Some soft-bodied sea creatures led on to animals with shells or inner skeletons. Fishes gave rise to amphibians; amphibians to reptiles; reptiles to the birds and mammals.

Through many generations tiny but accumulating changes in inherited characters altered organisms, better fitting them to feed, breed, and survive their enemies. So arose millions of species – twigs on a mighty tree of life. Related dead or living species form a genus. Related genera comprise a family. Related families belong to successively more major groupings – classes, phyla (animals) or divisions (plants), and kingdoms.

A fossil forming (left)
1 A dead fish on the sea bed.
2 Flesh rots, baring bones.
3 Mud or sand buries and preserves the bones.
4 Minerals harden bones now hidden by layered sediments.
5 Weather removes upraised rocks, exposing the fossil.

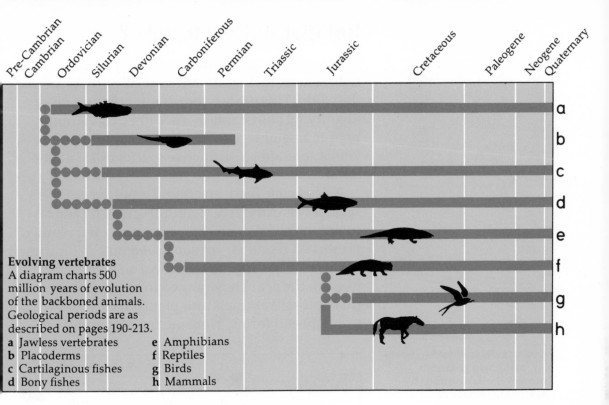

Evolving vertebrates
A diagram charts 500
million years of evolution
of the backboned animals.
Geological periods are as
described on pages 190-213.

a Jawless vertebrates
b Placoderms
c Cartilaginous fishes
d Bony fishes

e Amphibians
f Reptiles
g Birds
h Mammals

Most major groups are very old indeed. But there survives a mere fraction of the hundreds of millions of species evolving in the last 600 million years. New enemies, or harsh environmental change destroyed the rest. Fossil records in the rocks show waves of mass extinction, then explosive evolution as new organisms moved into habitats left empty by immense catastrophes.

Major extinctions (below)
This graph shows the rise and fall in numbers of named groups of living things in the last 600 million years or so. It hints at mass extinctions followed by explosive bursts of evolution.

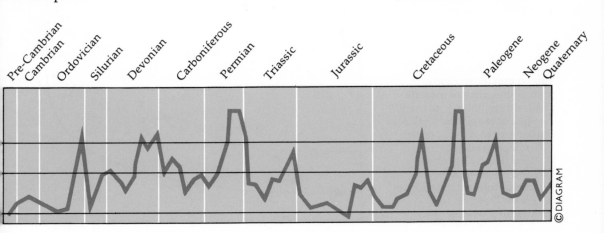

©DIAGRAM

175

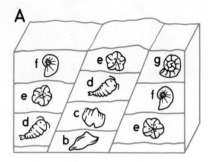

A

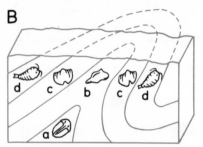

B

Relative dating: fossils 2

Fossils offer valuable aids to relatively dating sedimentary rock strata and correlating these around the world. This process, biostratigraphy, involves identifying faunal zones – rock strata containing unique assemblages of fossils. Geologists name each faunal zone after a distinctive species called a zone fossil.

A good zone fossil meets four requirements. Its species was extremely plentiful; spread far and fast (planktonic organisms are examples); left readily preserved remains; yet soon died out, so limiting its fossils to a few rock layers. Most such organisms lived in the sea. They ranged from sizeable (macrofossil) animals and plants to tiny (microfossil) forms. Here are four examples.

Fossils, faults, and folds (above)
Key fossils help geologists date rock layers disturbed by (**A**) faults or (**B**) folding. (For dates of rock systems see pages 180-181).
a Cambrian trilobite
b Ordovician crinoid ("sea lily")
c Silurian brachiopod ("lamp shell")
d Devonian eurypterid ("sea scorpion")
e Carboniferous blastoid (kin to starfish and sea urchin)
f Permian ceratite ammonoid
g Triassic ammonite

Faunal provinces (right)
Two groups of fossil trilobite (**a,b**) and graptolite (**c,d**) mark two faunal provinces - shelf seas flanking a pre-Atlantic ocean 500 million years ago.
A Land (Proto-Greenland)
B Land (Proto-Europe)
C Iapetus (pre-Atlantic) Ocean
D Pacific Province
E Atlantic Province

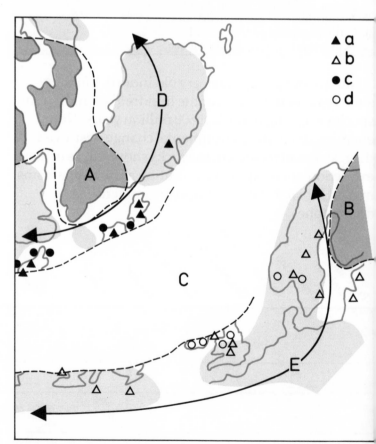

1 Trilobites ("three-lobed") were marine, segmented distant relatives of woodlice; zone macrofossils for rocks 590-510 million years old.

2 Ammonoids were cephalopod mollusks with coiled, flat, wrinkled shells; zone macrofossils for rocks 370-65 million years old.

3 Bivalves are headless mollusks with hinged, two-part shells; zone macrofossils for rocks 370-65 million years old.

4 Foraminiferans are tiny one-celled protozoan organisms drifting in the seas and forming limy shells pierced by tiny holes; zone microfossils for rocks 65-0 million years old.

Besides providing guides to evolution and the ages of rocks, fossil individuals and groups reveal how prehistoric living things behaved and the kinds of place and climate they inhabited.

There are limits to our knowledge. Most soft-bodied organisms left no fossil record. Relatively few land plants and animals were fossilized. Billions of fossils vanished when erosion wore away the rocks containing them, or these were baked or crushed by metamorphic change. Billions more are inaccessible. But new kinds of fossil are discovered every year.

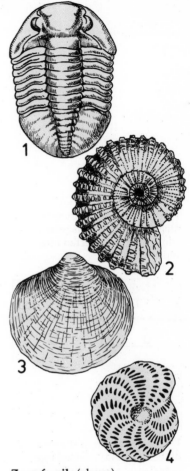

Zone fossils (above)
1 Trilobite
2 Ammonoid
3 Bivalve
4 Foraminiferan (much magnified)

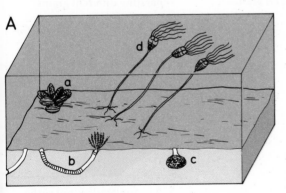

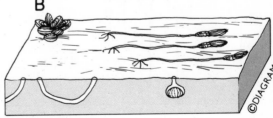

A Prehistoric community
a Shellfish
b Worm
c Sea urchin
d Crinoid (sea lily)

B Fossils as found
Undisturbed fossils (above) aided reconstruction of the seabed community (left).

177

Clocks in rocks

Chronometric dating gives approximate ages in years for the rocks. Some rare sediments are datable from annually added layers. Some rocks are dated fro the known rate of decay of a radioactive element into a more stable element. The more time that elapses, the less parent element remains and the more daughter element accumulates. So measuring the proportions of both elements within a rock reveals its age. For igneous rock this means how long ago its minerals crystallized; for sedimentary rock, when sedimentation produced certain minerals; for metamorphic rock, when heat drove daughter elements from the rock and "reset" the geological clo

Dating diabase (below)
Five illustrations depict six stages in dating a diabase (dolerite) dike injected into older rocks.

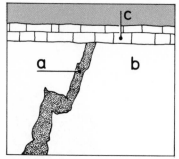

1 This cross section shows a diabase dike (**a**) injected into preexisting granite (**b**) before overlying sandstone (**c**) formed.

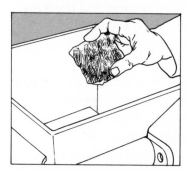

2 A lump of diabase is dropped into a crusher which grinds the rock until it breaks up into component mineral grains.

3 Froth flotation separates micas from the other minerals. Various methods serve for separating different grains.

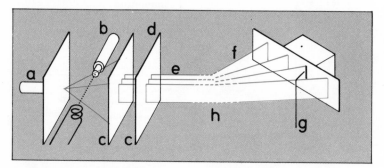

4 A mass spectrometer using a magnetic field separates and measures the amounts of isotopes of potassium and argon
a Gas inlet
b Electron beam
c Slits
d Ion accelerating voltage
e Ion beam
f Isotopes
g Detector slit
h Magnetic field

5 A computer printout of the isotope data enables calculation of the diabase's potassium-argon age. (Potassium is the parent element, and argon the daughter.)

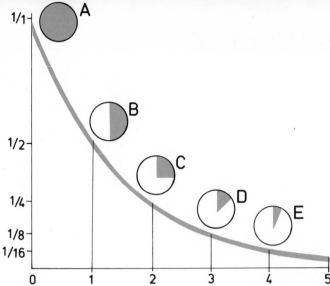

Radiometric dating (left)
By radiation potassium-40 loses half its mass every 1310 million years (one half life). Thus a sample's potassium-40 content can indicate its age.
A Original sample
B After 1310 million years (one half life) half remains.
C After 2620 million years (two half lives) one quarter remains.
D After 3930 million years (three half lives) one-eighth remains.
E After 5240 million years (four half lives) one-sixteenth remains.

Various elements and their isotopes (different forms of atom of a given element) give best results in different circumstances. Here are brief details of three radiometric techniques and one based on nuclear fission.

1 Potassium-argon dating exploits the decay of the potassium-40 isotope into argon-40. Applications: mainly igneous and metamorphic rocks of any age greater than about 1 million years, and sedimentary rocks containing the mineral glauconite.

2 Rubidium-strontium dating uses the decay of rubidium-87 to strontium-87. Applications: igneous and metamorphic rocks (except basic types), and sedimentary rocks containing the mineral illite. This method is often best for rocks more than 30 million years old.

3 Uranium-thorium-lead methods involve radioactive isotopes in uranium. Uranium-235 decays to lead-207; thorium-232 decays to lead-208. Applications: igneous intrusions, metamorphic rocks, and sediments containing zircon. These methods are best for rocks over 100 million years old.

4 Fission-track dating involves counting fission tracks produced in rock by splitting nuclei of uranium-238, whose nuclei split at a known and constant rate. The older the rock, the more fission tracks there are. Applications: many igneous and metamorphic rocks.

Fission-track dating
Fission tracks in a crystal are etched to make them show up, then counted under a microscope. The more tracks, the older the crystal.

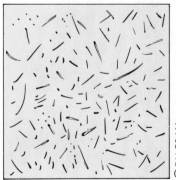

©DIAGRAM

The geological column

By studying strata, geologists built up the geological column – a full record of the Earth's crust's rocks laid down in sequence. This column is based on rock units, each comprising rock layers created in one geological period of time. In ascending order several of the time rock units called zones make up one stage; several stages form a series; several series build a system; several systems make an erathem.

At first, geologists assigned only relative dates to geological time intervals. Radiometric dating occasionally enables them to be much more exact. Yet dates are still approximate and less precise the farther back we go.

Grand Canyon (left)
From the rim of Arizona's Grand Canyon, spectators gaze down 6250ft (1900m) through more geological history than in any other place on Earth. Exposed rocks date from four billion to 250 million years ago. (See also page 211).

Geological column (right)
This shows time units of the Phanerozoic Eon. Start dates are in millions of years. Neogene and Paleogene are often combined as the Tertiary Period. The Carboniferous Period is divided into the Mississippian (360-320 million years ago) and Pennsylvanian (320-286 million years ago).

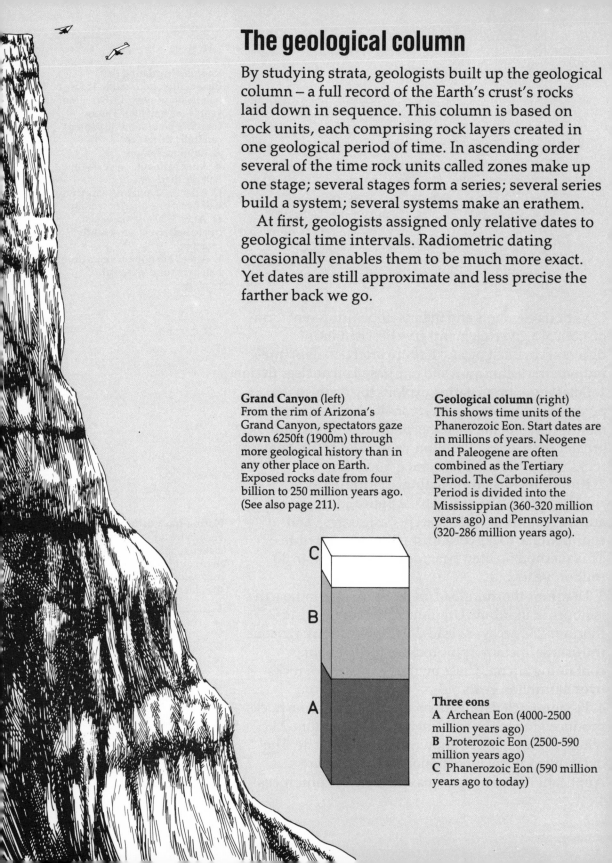

Three eons
A Archean Eon (4000-2500 million years ago)
B Proterozoic Eon (2500-590 million years ago)
C Phanerozoic Eon (590 million years ago to today)

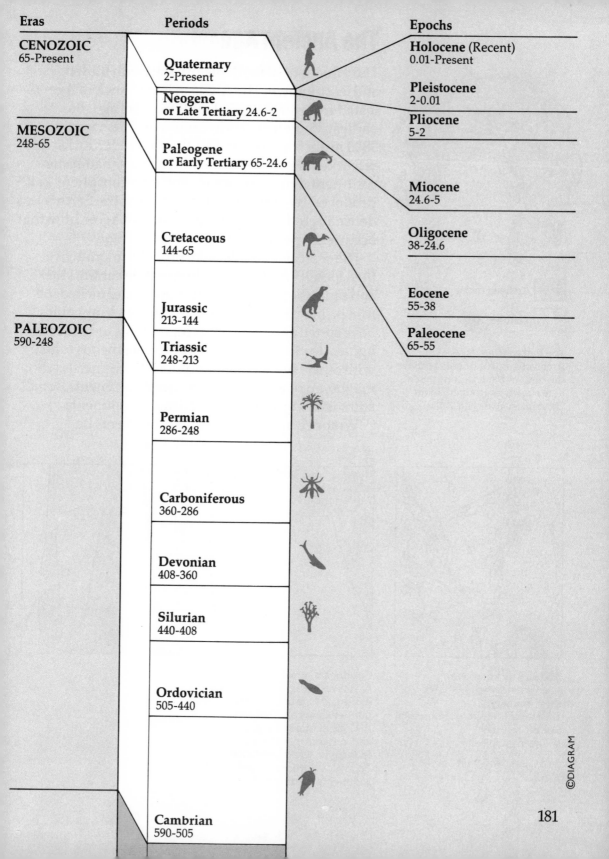

Eras	Periods	Epochs
CENOZOIC 65-Present	**Quaternary** 2-Present	**Holocene** (Recent) 0.01-Present
	Neogene or Late Tertiary 24.6-2	**Pleistocene** 2-0.01
MESOZOIC 248-65	**Paleogene** or Early Tertiary 65-24.6	**Pliocene** 5-2
	Cretaceous 144-65	**Miocene** 24.6-5
	Jurassic 213-144	**Oligocene** 38-24.6
PALEOZOIC 590-248	**Triassic** 248-213	**Eocene** 55-38
	Permian 286-248	**Paleocene** 65-55
	Carboniferous 360-286	
	Devonian 408-360	
	Silurian 440-408	
	Ordovician 505-440	
	Cambrian 590-505	

©DIAGRAM

181

The Ancient Age

The three eons of Earth history start with the longest and least-known. The Archean Eon ("Ancient Age") lasted from 4000 to 2500 million years ago. Its earliest-known surviving rocks on Earth are 3800 million years old. But they are older rocks reworked. Probably the world already had some continental rock, an ocean, and an atmosphere – all produced by sorting and resorting of the Earth's less dense ingredients. Geologists do not agree how that occurred – perhaps like this.

The crust and mantle were probably more active than today. Where two cooling mantle currents met and sank, they squashed, thickened, and melted the thin primeval crust above. Repeated melting could have sorted and resorted its ingredients until the lightest formed a scum of continental igneous rocks, with others highly metamorphosed into gneisses. So, maybe, appeared "granitoid" microcontinents, some still surviving as the ancient cores of continents.

Wrapped around these microcontinents lie

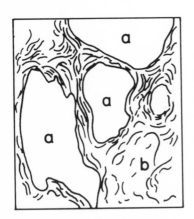

The first continents (above)
Archean minicontinents became the cores of modern continents, here shown as grouped about 250 million years ago.

Archean rocks

Younger rocks

Archean rocks (above)
A satellite-eye-view shows these features.
a Cratons (granitoid continental nuclei)
b Greenstone belts

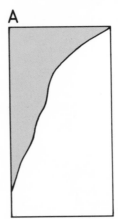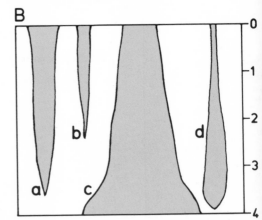

Continental growth (above)
A Increasing sediment volume in the last four billion years implies slow growth of land.
B Continental sediments (**a**) and volcanic rocks (**b**) increased in volume, while greenstone belt volcanic rocks (**c**) and sediments (**d**) diminished.

greenstone belts of lightly metamorphosed greenish dark volcanic rock, with shales and sandstones. The volcanic rock came perhaps from volcanoes spewing lava, ash, and gas from scores of hotspots in a weak, thin, early crust. Or they were island-arc volcanoes and the shales and sandstones formed from sediments washed off nearby microcontinents.

Perhaps volcanic steam that cooled and turned to rain filled early ocean basins and the rivers that eroded rocks, resorting their materials. Certainly volcanic gases formed an early atmosphere, rich in nitrogen and carbon dioxide.

Fossil organisms in the greenstone belts reveal that life appeared at least 3500 million years ago. Bacteria and blue-green algae were flourishing in shallow seas. Oxygen released each year by algae combined with iron, producing banded iron formations now mined around the world. But iron and other chemical sponges left little oxygen for adding to the atmosphere.

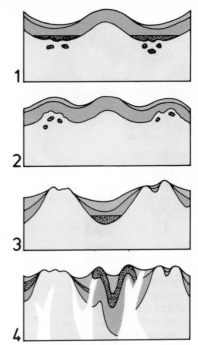

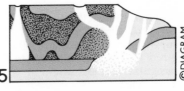

©DIAGRAM

Possible processes (above)
These stages might have formed Archean microcontinents.
1 Sediment from uplifted crust fills sags in oceanic crust.
2 Melting forms acid magma.
3 Subsidence and melting form alkaline magmas.
4 Folding yields granitic magmas and greenstone belts.
5 The cooled product is a sialic shield and mobile belt.

Stromatolites (left)
Blue-green algae formed intertidal pedestals like these 3500 million years ago.

The Age of Former Life

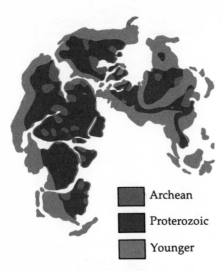

■	Archean
■	Proterozoic
■	Younger

Expanding continents (above)
Proterozoic rocks added new
land to the old, Archean,
minicontinents. Lands appear
as grouped 250 million years
ago.

A supercontinent (below)
A supposed mid-Proterozoic
supercontinent would have held
bits of the future Asia and these
recognizable land masses.
a Australia
b Antarctica
c India
d South America
e Africa
f Proto-North America
g Proto-Europe

About 2500 million years ago the first large continents
appeared, and extensive, shallow, offshore seas gave
new opportunities for living things. These changes
ushered in the second phase of Earth history. This
Proterozoic Eon ("Age of Former Life") spanned some
1900 million years, ending about 590 million years ago.

New "granitoid" and greenstone belts emerged, and
vast masses of volcanic rocks and sediments were
tacked on to Archean microcontinents.

Geologists detect three main construction phases,
starting 1900, 1200, and 700 million years ago. The first
produced the Wopmay Orogen – a long-since beveled
mountain belt in northern Canada. Such ancient sites
hold traces of old oceanic crust, island arcs, and
colliding continents. This shows that plate tectonics
was already molding continental and oceanic crust.
Alignments of magnetic particles in rocks prove
continents were drifting and ocean floor was rifting
and subducting by 1500 million years ago. Some
evidence suggests that major continents were even
stuck together at that time.

Meanwhile, the oceans and the atmosphere were
undergoing change. Salts washed off land gave the
sea its present saltiness. By about 2000 million years
ago, algal plants produced enough free oxygen for
some to start accumulating in the sea and
atmosphere. (Proof comes from compounds formed
in certain rocks.)

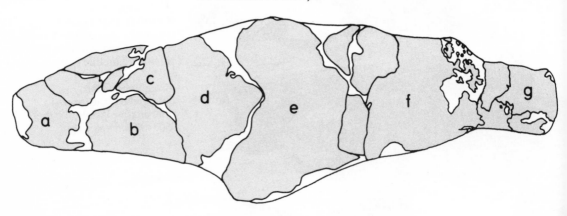

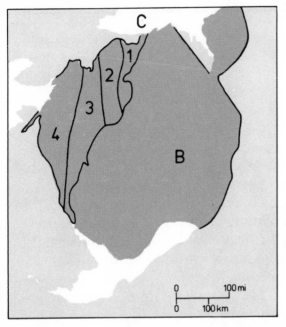

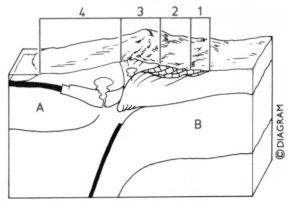

The Wopmay Orogen
Below: Colliding Bear Plate (**A**) and Slave Plate (**B**) crumpled intervening rock zones (**1-4**). Left: A map shows the same zones and the old Slave Plate all now in Canada south of the Arctic Ocean (**C**).

Atmospheric oxygen began to build an ozone shield protecting living things from the Sun's lethal ultraviolet radiation. New, complex kinds of water life evolved, able to exploit the energy in oxygen. Soon after 700 million years ago, soft corals, jellyfish, worms, and other soft-bodied animals were flourishing in shallow seas off continental shores.

Complex life (below)
Late Proterozoic seafloor organisms of Australia:
a Coelenterates (relatives of modern jellyfish)
b Segmented worms
c Arthropod (a shrimplike specimen)
d Algae
e Unknown

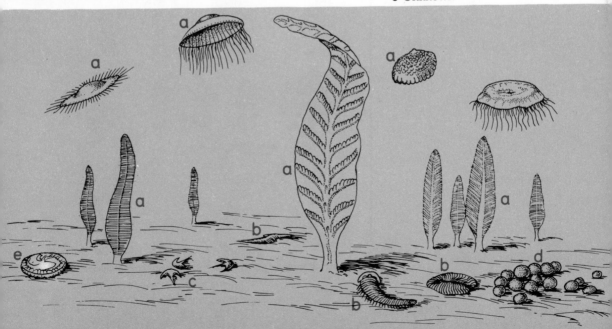

Chapter 11

THE LAST 600 MILLION YEARS

These pages trace major events of the Phanerozoic Eon – the last, shortest, and best recorded volume in Earth history. Era by era, period by period, we examine the vast changes that transformed our planet's continents, climates, and living things.

A section of the Wirksworth Cave, Derbyshire, England, and some of the fossils found there. (Engraving originally published in *The Iconographic Encyclopaedia of Science, Literature and Art* 1851)

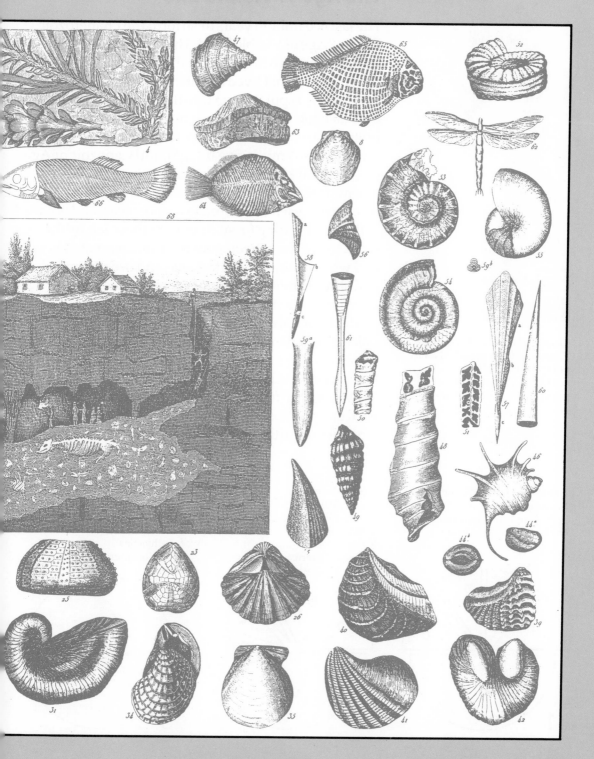

The Age of Visible Life

The last 600 million years or so form the Phanerozoic Eon or "Age of Visible Life." The Phanerozoic saw complex modern life forms in the making. Assemblages that lived at different times have led geologists to split the eon into three successive eras – the Paleozoic or Age of Ancient Life (about 590–248 million years ago), Mesozoic or Age of Middle Life (about 248–65 million years ago), and Cenozoic or Age of Recent Life (about 65–0 million years ago). Fossil assemblages differing in time and place help experts reconstruct each era's climates, lands, seas, and oceans.

By coordinating fossil clues with those left by the rocks themselves, geologists have reconstructed how and why continents drifted, seas and oceans spread and shrank, mountain ranges rose and were worn down, and ice sheets waxed and waned.

Much remains unknown. For instance paleomagnetism reveals the past north-south positions of the continents, but their longitudes (east-west locations) lie open to dispute.

Even so, the following pages can stress key events in each era's successive periods – chapters in the later story of the Earth.

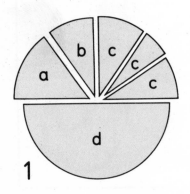

Phanerozoic

Proterozoic

Archean

Continents complete (above) Phanerozoic rocks tacked on to older rocks completed the continents, here shown as grouped 250 million years ago.

Continental drift (below) Segments show landmasses split, fused, and split again in the last 600 million years.

1 Early Paleozoic
2 Later Paleozoic
3 Latest Paleozoic
4 Mid Mesozoic
5 Cenozoic
a North America

b Europe
c Asia
d Gondwanaland
e Pangea
f Tethys Sea
g South America

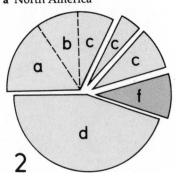

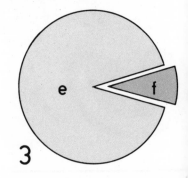

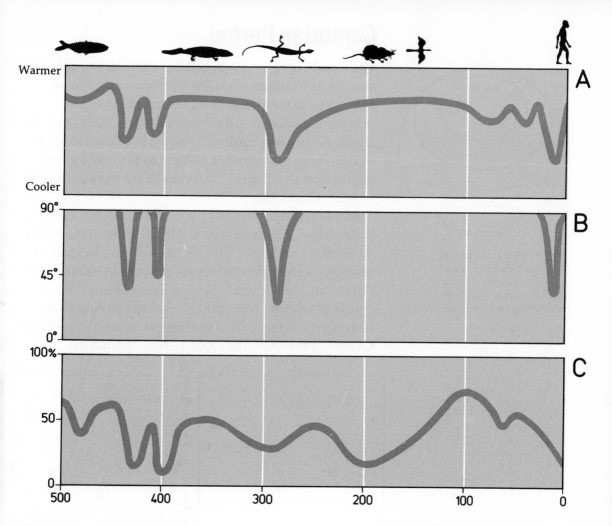

Warmer

A

Cooler

90°

B

45°

0°

100%

C

50

0

500   400   300   200   100   0

h Eurasia
i Antarctica/Australia
j Africa
k India
l Antarctica
m Australia

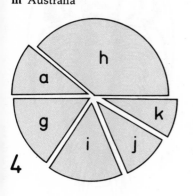

4

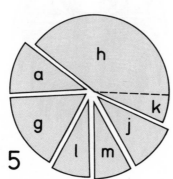

5

Climatic changes (above)
Three graphs show climatic and
related changes in the last 500
million years.
A Likely rise and fall in world
temperatures
B Glaciers' advance toward and
retreat from the equator, at
latitude 0 degrees
C Spread and shrinkage of
continental seas

©DIAGRAM

189

Cambrian Period

The Cambrian Period (about 590–505 million years ago) takes its name from the Latin word for Wales. Here geologists first studied Cambrian fossils – notably the first abundant animals with skeletons and shells. All life was limited to water. It teemed in shallow seas invading continents as ice sheets melted about 600 million years ago. Cambrian times were generally warmer than today.

Most continental lands probably lay on or close to the equator. South America, Africa, India, Antarctica, Australia, and bits of Asia were evidently welded in one southern supercontinent – Gondwanaland – with Africa "upside down." Among lesser chunks of continental crust were the cores of North America, Greenland, Europe, and northwest Africa. The Iapetus Ocean – a pre-Atlantic Ocean – had opened up

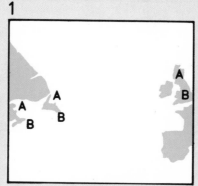

Cambrian world (above)
Lands might have been arranged like this. Lines show the Equator and latitudes 30 and 60 degrees north and south of it.

1

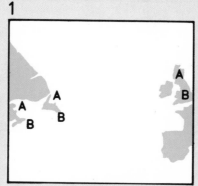

2

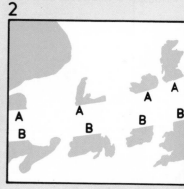

A trilobite tale (right)
Two Cambrian trilobites (right) posed a geologic puzzle. Fossil *Olenellus* (**A**) occurs as seen in Map 1 – in northern parts of Nova Scotia, Newfoundland, and the British Isles (shown closer than they really are). Fossil *Paradoxides* (**B**) occurs farther south in all three places. Map 2 shows why. In Cambrian times each place was two areas of shallow sea floor separated by the Iapetus Ocean. This deep-sea barrier kept both kinds of trilobite apart.

A

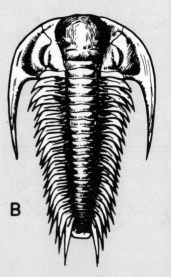

B

190

between these once-fused lands. Within this ocean lay Avalonia, an archipelago whose rocks today lie scattered from the Carolinas north through Newfoundland to parts of Ireland and Wales.

By 570 million years ago mountains had sprouted in the Avalonian orogeny when a slab of land struck eastern North America to form New England. Elsewhere, the Andes mountains had begun to grow, and volcanoes spewed vast sheets of lava over parts of north and west Australia.

Sands washed off land into shallow seas provided raw materials for sandstones. Animals with shells became a source of carbonates, and dolomite began to form. Other Cambrian sediments include blue clays that still survive in Russia.

Cambrian rocks exposed
Cambrian limestone overlying Cambrian quartzite form these high cliffs in Canada's Banff National Park. The cliffs' materials were laid down as immensely thick deposits in shallow water off what was then the northwest rim of proto-North America.

Ordovician Period

Rocks from this time (about 505–440 million years ago) were first studied in Wales; an early Welsh tribe, the Ordovices, inspired the period's name.

Subduction brought slow shrinkage of the Iapetus Ocean and a closing up of its flanking continental cores: Laurentia (proto-North America and Greenland), Baltica (proto-Europe) and northwest Africa. Early on, subducting or colliding crustal blocks deformed rocks of the future Scottish Highlands. Later, the Taconic orogeny forced up the Green Mountains of Vermont. Volcanic rocks appeared in Scandinavia and Greenland.

Although large slabs of continental crust lay close to the equator, part of Gondwanaland moved deep into Antarctic latitudes. Indeed northwest Africa lay astride the South Pole. Late Ordovician ice-scoured rocks and iceborne debris show that ice sheets covered northwest Africa and nearby parts of South America.

Ordovician world (above) Lands might have been arranged like this. Lines show the Equator and latitudes 30 and 60 degrees north and south.

Future mountains (below) Rocks eroded from Gondwanaland rimmed that land with belts of sediments in downwarped crustal troughs – a later source of mountain chains and ranges.

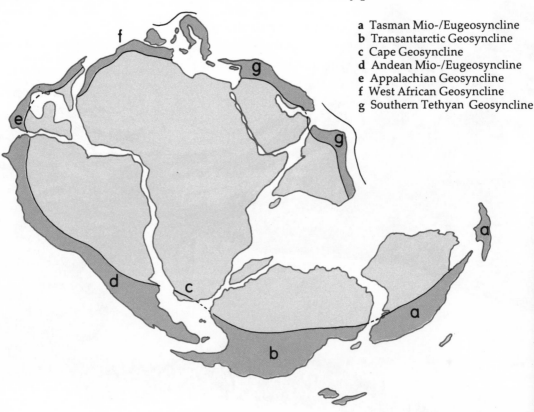

a Tasman Mio-/Eugeosyncline
b Transantarctic Geosyncline
c Cape Geosyncline
d Andean Mio-/Eugeosyncline
e Appalachian Geosyncline
f West African Geosyncline
g Southern Tethyan Geosyncline

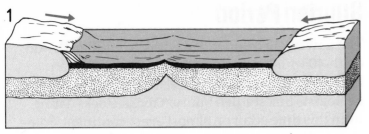

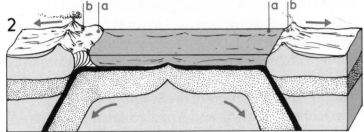

At their maximum extent the ice sheets locked up
much of the world's water, and shallow continental
seas withdrew. But from time to time, warm, shallow,
salty water invaded low lands including parts of
proto-North America. Here lived trilobites, early
corals, and many more invertebrates. The small
colonial organisms called graptolites left fossils used
for correlating the ages of Ordovician rocks from
different areas. There were early jawless fishes, too.

Ordovician sediments up to about 23,000ft (7000m)
thick formed in continental seas and offshore waters.
North America, Europe, and north Australia all
accumulated beds of limestone, dolomite, and coral.

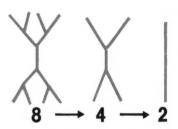

Evolving graptolites (above)
Ordovician times saw a
reduction in the number of
branches formed by colonies of
tiny sea creatures, called
graptolites

Ordovician rocks exposed
The shales and sandstones
above this road at South Africa's
Cape of Good Hope contain late
Ordovician fossils and pebbles
scratched by Ordovician
glaciers.

Silurian Period

The Lower Paleozoic ended with the Silurian Period (440–408 million years ago). The Silures were an ancient British tribe of the Welsh border region where geologists first studied Silurian rocks. Rocks dating from this time occur on almost every continent. Some contain the oldest fossil land plants and animals.

As the pre-Atlantic Iapetus Ocean shrank, proto-North America with Greenland were beginning to collide with proto-Europe. The so-called Caledonian orogeny crumpled up the edges of their plates, pushing up forerunners of the Scandinavian, Caledonian (Scottish), and Appalachian mountains – a mighty chain of peaks that would extend in time from Scandinavia through the British Isles and Greenland to New York. Farther west, North America ended at eastern Nevada and Idaho. But as the continent began to override the oceanic plate beyond, new land would stick on to its western rim.

Some experts think Asia was mainly three ocean-isolated blocks – Siberia, China, and part of South-East Asia. But the first two were closing by Silurian times. Indeed perhaps all northern continental slabs had almost fused to form a northern supercontinent, Laurasia.

Silurian world (above) Lands might have been arranged like this. Lines show the Equator and latitudes 30 and 60 degrees north and south.

Alternative view (below) Some experts think today's northern continents remained still largely unassembled.
a Bits of North America
b Bits of Europe
c Bits of Asia (perhaps even more than shown)
d Southern continents with bits of northern ones

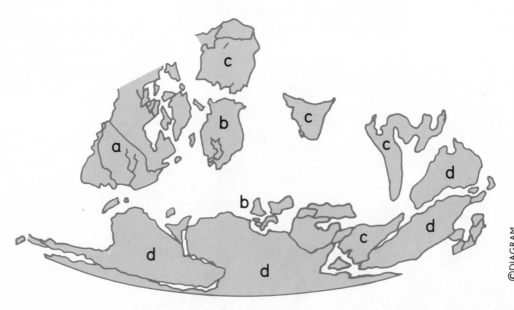

©DIAGRAM

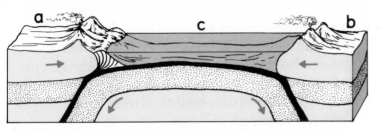

A closing ocean (left)
Proto-North America (**a**) and
proto-Europe (**b**) advance on
one another, closing the Iapetus
Ocean (**c**) and starting the
Caledonian orogeny.

In the southern supercontinent, Gondwanaland,
Africa and South America were drifting north while
Antarctica and Australia still headed south.

Melting southern ice sheets flooded continents with
shallow seas. Off proto-North America the sea floor
gained thick sheets of sands and gravels, the eroded
ruins of high mountains raised in Ordovician times.
Other sediments produced rich oil reserves in what is
now the Sahara Desert. Widespread reefs marked the
spread of (solitary) corals, and evaporites accumulated
on the arid western coasts of continents.

Silurian life (left)
1 *Cystiphyllum,* a solitary coral
2 *Baragwanathia,* a lycopod – an
early land plant
3 *Palaeophonus,* a scorpion –
one of the first land animals

Silurian rocks exposed
These Silurian grits on the west
Welsh coast formed horizontal
layers until tilted by the
Caledonian orogeny.

Devonian Period

The Devonian Period takes its name from Devon, England, where its shales, slates and Old Red Sandstone were laid down about 408–360 million years ago. But every continent has rocks dating from this first phase of the Upper Paleozoic. Devonian deposits include widespread coral reefs, and rich oil reserves in Canada and Texas.

During the Devonian, subducting oceanic crust and colliding northern continents entirely closed the northern Iapetus Ocean. The Acadian orogeny uplifted much of northeast North America while the Caledonian orogeny was still affecting Europe. Eastern North America, Greenland, and western Europe fused to form a so-called Old Red Continent. Its Old Red Sandstone rocks are formed from the eroded fragments of the huge mountain chain thrown up by the collision. Fossils in such rocks include freshwater fishes and the first amphibians, whose Greenland home then straddled the equator. Devonian rocks also hold fossil remnants of the world's first forests.

Devonian world (above) Lands might have been arranged like this. Lines show the Equator and latitudes 30 and 60 degrees north and south.

Devonian life (below)
1 Brachiopods ("lamp shells")
2 *Hemicyclaspis,* a jawless fish, shown actual size
3 *Ichthyostega,* an early amphibian 3ft 3in (1m) long

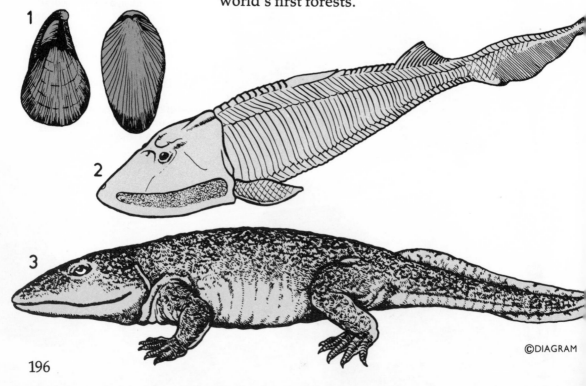

©DIAGRAM

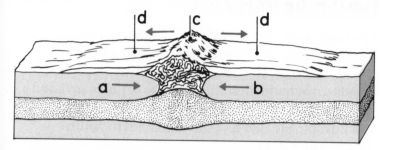

Caledonian-Acadian orogeny
Left: The colliding plates of proto-North America (**a**) and proto-Europe (**b**) closed the Iapetus Ocean, crumpling intervening rocks (**c**) into a chain of peaks from Scandinavia to New York. Their eroded remnants formed Old Red Sandstone (**d**).

Now, Gondwanaland was moving north, and pushing minicontinents ahead of it. Only a narrow, shrinking sea, the Tethys, separated South America and Africa from North America and Europe. Meanwhile the ocean separating Russia from Siberia was evidently closing. Fresh collisions were inevitable. One theory holds that most of western Europe was created when the minicontinent Armorica slammed into Baltica, the proto-European continent. In Devonian times, this impact was foreshadowed by heavings that began pushing up the Hercynian mountain belt whose remnants include the Armorican Massif, Vosges, and Black Forest.

Devonian rocks exposed
Gently dipping Upper Old Red Sandstone beds overlie more steeply dipping Lower Old Red Sandstone here in eastern Scotland.

Carboniferous Period

The Carboniferous Period (360–286 million years ago) takes its name from thick coal-producing carbon layers. These are the remains of swampy tropical forests drowned when shallow seas at times invaded a vast low-lying tract embracing much of North America and Europe. Smaller forests flourished in South America and Asia.

In North America this period is split in two. The Mississippian (360–320 million years ago) saw limestones laid down by a shallow sea covering the Mississippi region. The Pennsylvanian (320–286 million years ago) is named from coal measures formed in Pennsylvania about 320–286 million years ago. Coal forests then flourishing in Nova Scotia contained the first known reptiles.

The Carboniferous saw North America and Europe colliding with the northern edges of Gondwanaland – the part containing South America and Africa. By about 350 million years ago, this process was fusing

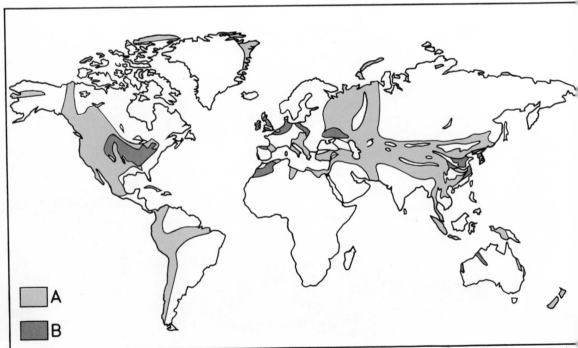

Carboniferous world (above)
Lands maybe reached these positions. Lines mark the Equator and latitudes 30 and 60 degrees north and south.

Land and sea (below)
A modern map shows (**A**) land under sea and (**B**) main lowlands often drowned by sea about 300 million years ago.

©DIAGR

northern and southern continents into a single landmass, called Pangea.

The slow collisions forced up mountain ranges. About 300 million years ago South America seemingly struck Texas and Oklahoma, pushing up the Ouachita Mountains. Later, South America or Africa smashed into southeast North America. This gave the Alleghenian orogeny, crumpling up the southern Appalachians. Meanwhile South America or Africa colliding with the south of Europe destroyed the intervening sea, and continued raising the Hercynian mountains, whose eroded roots still run from southern Ireland to Bohemia. Such impacts also generated mountains in Gondwanaland.

As drifting carried parts of southern continents across the South Pole, ice sheets again smothered regions of the Southern Hemisphere. By late Carboniferous times, ice covered all Antarctica, parts of Australia, and much of southern South America, Africa, and India.

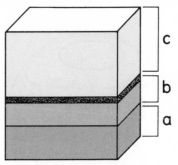

A sediment sequence (above) Non-marine sandstone (**a**), transitional sediments (**b**) including coal, and marine sediments (**c**) formed in recurring sequence in Late Pennsylvanian times.

Carboniferous life (below)
1 Conodont animal, a tiny sea creature, enlarged. Conodonts or "cone-teeth" supported soft tissue in the gut area.
2 *Meganeura*, a giant "proto-dragonfly"
3 *Hylonomus*, an early reptile, about 3ft 3in (1m) long.

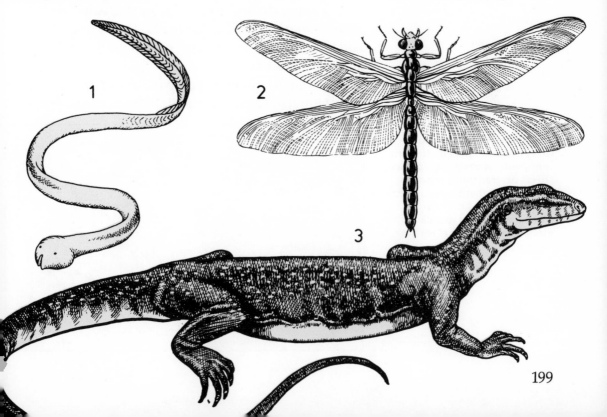

Permian Period

The Permian Period (about 286–248 million years ago) is named after rocks from the old province of Perm in Russia's Ural Mountains.

This wall between northern Europe and Asia sprang up in Permian times when the Siberian plate collided with eastern Russia. Meanwhile other "pre-Asian" plates were quite likely docking now, deforming rocks of intervening mobile belts in forging most of the rest of Asia.

Elsewhere, Africa's (or South America's) collision with Europe's southern underbelly had buckled up the mountains of Europe's Hercynian mobile belt. Farther west, Africa's (or South America's) collision with southeast North America went on crumpling up the southern Appalachians. On several continents collisions cracked open the crust, releasing basalt lavas.

Finally all continents lay jammed together as the supercontinent Pangea, surrounded by the single mighty Panthalassa Ocean.

As Pangea drifted north, glaciers retreated south in South America, Africa, and India. They gripped Antarctica as that landmass crossed the South Pole, also areas of Australia.

With much water locked up in ice and uplift of some continental masses, continental seas began to drain

Permian world (above)
Lands might have looked like this, omitting (much shrunken) continental seas. Lines show the Equator and latitudes 30 and 60 degrees north and south.

Permian ice sheets
This map of Early Permian Gondwanaland (fused southern continents) suggests moving ice sheets covered much of South America, southern Africa, India, Antarctica, and southern Australia. Scoured rocks and glacial deposits hint at ice flow and extent.

away. Large tracts of northern continents experienced a dry, continental climate with deserts that contained evaporites. Iron-rich minerals strongly oxidized in warm conditions produced the vivid rusty red beds typical of sedimentary rocks laid down in these conditions.

As the Permian Period (and Paleozoic Era) closed, the loss of many continental seas, amongst other factors, helped produce the greatest-ever mass extinctions in the fossil record.

Permian life
1 *Medlicottia*, an ammonoid
2 *Dimetrodon*, a flesh-eating reptile 11ft 6in (3.5m) long
3 Conifer, a tree bearing seeds in cones

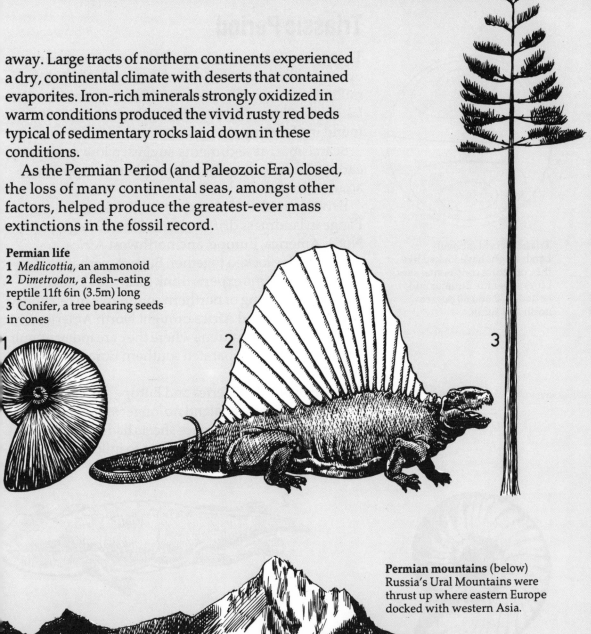

Permian mountains (below) Russia's Ural Mountains were thrust up where eastern Europe docked with western Asia.

Triassic Period

The Mesozoic Era, often called the Age of Dinosaurs, opened with this period dating from about 248–213 million years ago. The name "Triassic" comes from the Latin *trias* ("three") derived from three rock layers found in Germany.

Scarce marine sediments suggest a low sea level early in Triassic times, but red beds and evaporites accumulated on the land.

Between mid-Permian and mid-Triassic, the Pangean landmass drifted north about 30 degrees. North America, Europe, and northwest Africa seemingly lay locked together. But not, perhaps, immovably. Some experts think a 2200mi (3500km) east-west shearing of northern continents in relation to South America and Africa brought North America and Europe closer to positions where they are today. A gulf – the Tethys Sea – separated southern Eurasia from Afro-India.

Much of North America and Europe still lay inside the tropics. Gondwanaland no longer straddled the South Pole, and southern ice sheets had all melted. World climates ranged from warm to mild and deserts were extensive.

Triassic world (above)
Lands might have looked like this, omitting continental seas. Lines show the Equator and latitudes 30 and 60 degrees north and south.

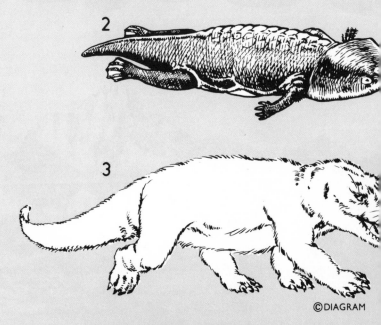

Triassic life
1 *Tropites*, a ceratite ammonoid
2 *Gerrothorax*, an amphibian 3ft 3in (1m) long
3 *Cynognathus*, an advanced mammal-like reptile 5ft (1.5m) long

©DIAGRAM

202

Pangea now showed signs of breaking up. Here and there, rising plumes of matter in the mantle domed the crust above until it split, creating block faults leaking lava. Indeed such rifts dated back to Carboniferous times in Scotland's Midland Valley and a Permian rift opened up in Norway. Triassic rifts affected west and central Europe, eastern North America, and north-west Africa. But Pangea's true destruction lay ahead.

Cracked supercontinent
The Palisades along New York's Hudson River are a Triassic or early Jurassic sill 400ft (120m) high. Its molten diabase (dolerite) rock rose through a crustal rift foreshadowing the break-up of Pangea.

Jurassic Period

The Jurassic Period (about 213–144 million years) takes its name from fossil-bearing limestone rocks formed in a sea but later raised as part of Europe's Jura Mountains. World climates were now mostly warm, and lands largely low, with old Paleozoic mountains worn down into stubs. Dinosaurs could have wandered overland across the world they shared with early birds and mammals. But the continental crust was growing restless.

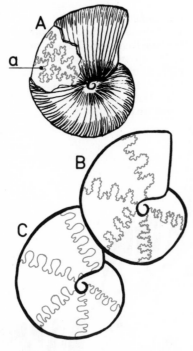

Jurassic world (above)
Lands might have looked like this, omitting continental seas

Fossil ammonite (above)
A *Phylloceras*: a Jurassic ammonite, an ammonoid with distinctive shell sutures (**a**).
B Ammonite shell (minus covering) – all sutures frilled
C Earlier ammonoid – not all sutures frilled

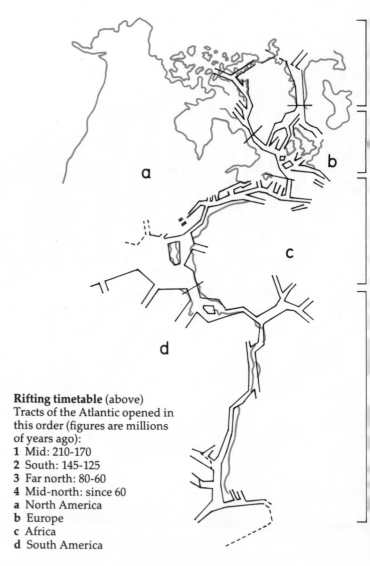

Rifting timetable (above)
Tracts of the Atlantic opened in this order (figures are millions of years ago):
1 Mid: 210-170
2 South: 145-125
3 Far north: 80-60
4 Mid-north: since 60
a North America
b Europe
c Africa
d South America

Pangea had started breaking up into the continents we know today. Here and there crust domed then split, creating triple-junction rifts. Later, linked rifts formed a spreading ridge that opened up the central part of the Atlantic Ocean, divorcing eastern North America from northwest Africa. Bits of the original continents became transposed; North America probably gained Florida from Africa. (Elsewhere, Asia would eventually gain Siberia's eastern tip from North America.)

Meanwhile, rifting was separating Africa/South America from Antarctica/Australia, although the Indian subcontinent was probably still stuck to eastern Africa. As cracks appeared vast flows of molten basalt welled up from southern Africa, through Antarctic mountains to Tasmania. More basalt flows show Australia preparing to cast off from Antarctica.

About 145 million years ago, Africa pushed east against southern Europe, shedding crustal chunks. These minicontinents eventually formed parts of Spain, Italy, Greece, Turkey, Iran, and Arabia. Meanwhile Eurasia had been fusing with Tibet.

Moving west to override the ocean floor, western North America produced three mountain-building episodes. Volcanoes sprouted as far south as the central Andes. Mighty blobs of molten granite bobbed up, melting solid rocks above. And slabs of crust – some possibly from Asia – got jammed against the western rim of North America.

A limestone house (above)
Jurassic limestone provided walls for this fine old English Cotswold manor house.

Life on land (below)
1 Bennettitalian, a plant with palm-like fronds
2 Sauropod, a giant plant-eating type of dinosaur with four limbs, long neck, and long tail
3 *Archaeopteryx*, the first known bird

© DIAGRAM

Cretaceous Period

The Mesozoic Era closed with the Cretaceous Period (144–65 million years ago). Its name comes from the Latin *creta*, meaning "chalk." Thick chalk deposits formed in shallow seas invading Europe, North America, and west Australia. Other deposits include 60 per cent of today's known oil reserves.

Pangea was now fragmenting into (northern) Laurasia and (southern) Gondwanaland, and both these supercontinents were also cracking up.

Early on, a spreading rift opened up the South Atlantic Ocean, driving South America and Africa apart. Much later, rifting separated Scandinavia from north Greenland, compressing and uplifting part of Siberia to raise the Verkhoyansk Mountains. But land still linked North America and Europe via south Greenland and the British Isles.

Farther south, the Bay of Biscay gaped open as north Spain pivoted away from west France. South of Europe, Africa moved east, forcing "Adriatica" against the Balkan plate, then moving west again. The Mediterranean was forming as a pinched-off portion of the Tethys Sea.

Dramatic changes added land and mountains to North America. Early on, part projected far toward the North Pole. Then the north split open and the north-west pivoted west, reacting with the Pacific Plate to

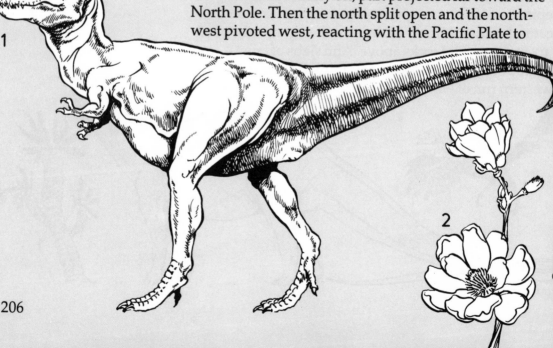

Cretaceous world (above)
Lands might have looked like this, omitting land bridges and continental seas.

Cretaceous life (below)
1 *Tyrannosaurus*, an immense flesh-eating dinosaur
2 Magnolia, a flowering plant

1

2

©DIAGRAM

206

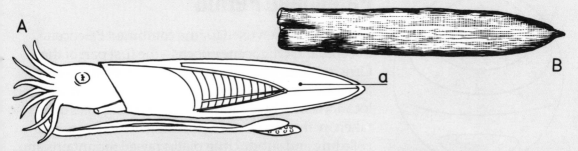

ruck up rocks into the Brooks Range of north Alaska. Arctic islands also probably rotated to where they lie today.

Western North America overriding oceanic plates continued spawning a great inland arc of batholiths and Andean-type volcanoes. Late in the Cretaceous, North America's westward drift accelerated, speeding up subduction of Pacific crust. This crumpled up the Rockies, exposing metals manufactured deep down in the crust and realigning rivers of the continent.

Meanwhile, in the Southern Hemisphere, India had cast adrift from East Africa, and New Zealand had most likely torn free from Australia.

Cretaceous climates remained chiefly warm or mild. Flowering plants began to spread. But the Mesozoic ended with the mass death of the dinosaurs and many other creatures. Impact of a huge asteroid might have caused climatic changes leading to this mass extinction.

Belemnite (above)
A Reconstruction of a belemnite up to 31in (80cm) long. A calcite rod (**a**), the guard, provided internal support for this Mesozoic kin of the squid and octopus.
B Fossilized belemnite guard. Fossil guards abound in some marine Mesozoic rocks.

Cretaceous cliffs (below)
Billions of shells of micro-organisms helped build these chalk cliffs at Beer Cove in southwest England.

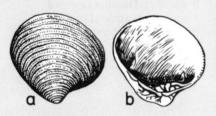

Paleogene Period

This term is often used for the combined Paleocene, Eocene, and Oligocene epochs – the first part of the Cenozoic Era. The Paleogene (65–24.6 million years ago) saw continents taking on their present shapes and locations, and birds and mammals filling roles once taken by the dinosaurs. Spreading ocean floors and colliding and subducting plates raised mountains and remade the map.

Shallow continental seas withdrew at first. Later, for a time sea invaded parts of Africa, Australia, and Siberia.

Western North America was thrusting east and overriding cool oceanic crust. This warmed up deep down and expanded, lifting all western North America by 30 million years ago. Meanwhile the Rocky Mountains and Colorado Plateau were evolving. Volcanoes spewed ash, and lowland sediments formed vast oil-shale deposits. In the northeast, by 45 million years ago the widening Atlantic Ocean had parted North America from Europe. To the south, immensely thick sediments pushed the Mississippi Delta out into the Gulf of Mexico, and North and South America separated.

Paleogene world (above)
The world roughly looked like this as India neared Asia in the Paleogene (alias early Tertiary) Period.

Familiar fossil (above)
Outer (**a**) and inner (**b**) views of one valve of a Venus shell. Dating from the Oligocene, this form of bivalve mollusk still flourishes.

Paleogene life (right)
1 *Uintatherium*, a rhinoceros-sized hoofed herbivore from Eocene North America
2 *Diatryma*, a giant ground bird of Eocene North America

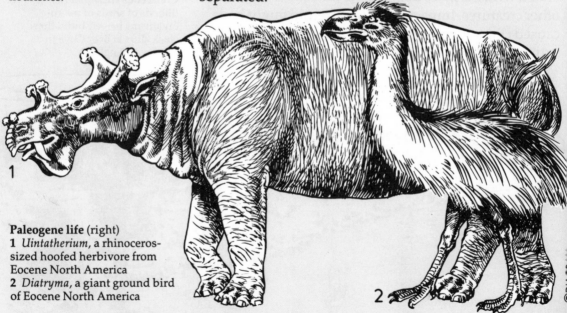

About 45 million years ago, Africa thrust north,
driving lithospheric platelets into Europe. Island
Iberia struck France and crumpled up the Pyrenees.
Farther east, the Adriatic plate overriding Europe's rim
began pushing up the Alps. About 30 million years
ago, part of France pivoted eastward, shoveling seabed
sediments ashore on Italy to form the Apennines.
Africa eventually added Sicily and the toe of Italy.
Much of southeast Europe formed when two small
Balkan plates struck southwest Russia.

In Africa itself great tracts warped up before
splitting to release vast lava flows and open up the Red
Sea rift.

By 40 million years ago, northward drifting India
had struck Siberia and the small Kazakstan and Tarim
plates to its south. The impact concertinaed the
collision zones and began to raise the Himalayas.

By 30 million years ago Antarctica formed an
Antarctic island continent. Chilled by its location, it
indirectly helped to lower temperatures worldwide.

Paleogene peak (right)
Mountaineers scale a peak in the
Rockies – ice-sculpted as crustal
heaving forced it high into the
chilly upper air.

Neogene Period

The Neogene Period (24.6–2 million years ago) comprises the combined Miocene and Pliocene epochs. Continents had almost reached their present places, and crashing plates were pushing up great modern mountain ranges.

By Mid-Cenozoic times, subducting seabed was forging island arcs around the west and north Pacific Ocean. Pacific area plate movements also cast adrift whole strips of continental crust. Japan probably split away from mainland Asia. The peninsula of Baja California was torn from mainland Mexico, and rode north-west along with California west of the San Andreas Fault.

Indeed faulting or volcanic eruptions racked western North America from Alaska south to Mexico. The Coast and Cascade ranges sprouted. Fissure flows built a vast basalt plateau in Oregon, Washington, and Idaho. Block faults from Nevada to Mexico formed the parallel ranges and valleys of the Basin and Range Province. But rising of the Rockies and Appalachians suggests upwarping of the whole continent. Rejuvenated mountain rivers eroded sharply downwards; the Colorado River was now carving out the Grand Canyon. Farther south a land bridge rejoined South and North America, and volcanic peaks were rising in the Andes.

Neogene world (above)
Lands reached present positions by the end of the Neogene (late Tertiary) Period.

Neogene fossil (above)
Cypraea, the cowrie, is a mollusk whose oldest fossils crop up in rocks formed on the floors of Miocene seas.

The shrinking Tethys (right)
A map shows Miocene remnants of the once mighty Tethys Sea and (**a- c**) outlines of their modern relics, cut off by shifting crustal plates. The Mediterranean (**a**) once linked the Atlantic and Indian oceans, and the Black Sea (**b**) joined the Caspian (**c**).

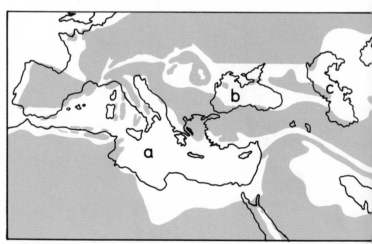

©DIAGRA

Out in the Pacific Ocean, volcanoes were spawning the Hawaiian Islands chain. Australia moving north collided with the Pacific Plate. This forced up mountains in New Guinea, and island stepping stones between Australia and Asia.

Meanwhile Africa's impact with Europe was manufacturing the Alps and the Carpathian and Atlas mountains. The Red Sea rift prised Africa away from Arabia and volcanic peaks arose along the African rift system. Arabia and Iran crashed into Asia to create the Taurus and Zagros mountains. Farther east, India's advance thrust up the Himalayas.

By 10 million years ago, Turkish and Arabian plates moving north had cut off the Mediterranean from the Indian Ocean, and Morocco had hit Spain. The thus isolated Mediterranean dried up and was refilled several times, leaving salt beds 20,000ft (6,000m) thick. Then, about five million years ago, a giant Atlantic waterfall burst in at Gibraltar and the sea refilled. Both polar regions now had ice caps.

Evolving horses (above)
1 *Miohippus*, a small three-toed horse – early Miocene
2 *Merychippus*, larger, and walking on each middle toe – Miocene
3 *Pliohippus*, the first one-toed horse – Pliocene

Rising land (below)
A section through one side of the Grand Canyon (**a**) shows multilayered and faulted ancient rocks laid bare as the Colorado River (**b**) gnawed down through a rising plateau (**c**).

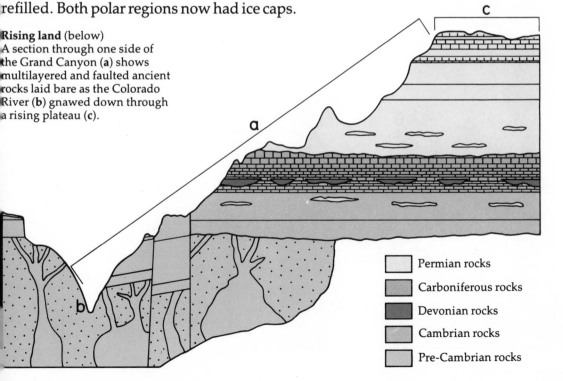

Permian rocks

Carboniferous rocks

Devonian rocks

Cambrian rocks

Pre-Cambrian rocks

Quaternary Period

This followed the Tertiary Period (alias Paleogene and Neogene periods). The Quaternary began about two million years ago. It embraced the Pleistocene or "Ice Age" Epoch and the mild Holocene or Recent Epoch in which we live today.

Pleistocene ice sheets smothered vast northern tracts, and glaciers filled mountain valleys worldwide. As climate fluctuated, ice repeatedly advanced and retreated – scouring valleys, damming lakes, rerouting rivers, and dumping debris over much of northern North America and Europe.

In intense glaciations sea level fell by as much as 330ft (100m). Meltwater torrents carved canyons in the rims of continental shelves. Land bridges joined Alaska and Siberia, mainland Asia and Indonesia, New Guinea and Australia, the British Isles and mainland Europe.

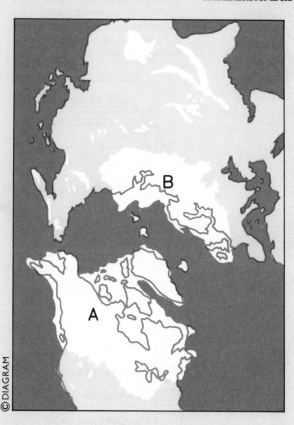

Quaternary world (above) Continents have shifted little in the last two million years, though levels of the land and sea have changed.

Lands under ice (left) Pleistocene ice sheets sometimes covered these labeled parts of (**A**) North America and (**B**) Eurasia.

Clues to cold
Below: This (much enlarged) foraminiferan coils right in warm water, left in cold.
Right: proportions of left- and right-coilers from 8m (26ft) seabed cores hint at past climatic changes.

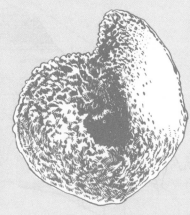

Volcanic power (right)
In 1980 Mt. St. Helens proved
explosively that a Cascades
Range volcano can still be
dangerously active.

Most of the northern ice sheets melted about 10,000
years ago. Sea levels rose, drowning the old canyons
and land bridges. The Black Sea and Mediterranean
were reunited. With the Caspian, both had once
formed part of the prehistoric Tethys Sea until cut off
by continental plates advancing from the south. As ice
melted, crust once weighed down by ice bobbed up; it
rises still in parts of Canada and Scandinavia.

Meanwhile great subterranean forces were at work.
Earthquakes, volcanic eruptions, oceanic trenches, and
high peaks show where lithospheric plates still
separate, collide, or grind against each other. For
shifting plates are still raising mountains and forcing
oceanic crust into the mantle. Thus Cascades Range
volcanoes sprout above the Farallon oceanic plate,
melting as it burrows under Oregon and Washington.
And earthquakes shake the San Andreas Fault as the
Pacific Plate bears western California north.

Africa pushing under Europe has formed a volcanic
island arc in the Aegean and raised Italy's volcanic
Etna and Vesuvius. A mighty rift broadens from the
Dead Sea to the Gulf of Aden. Arabia thrusts against
Iran. And farther east the Himalayas are still growing.

Man evolved, but wildlife wanes as our inventive
species competes with native plants and animals for
food and living space.

Tomorrow's World

Future world (above)
Lands and oceans might be grouped like this 50 million years from now.

Short- and long-term changes – some brought about by man – will drastically affect our planet's atmosphere, continents, and oceans.

By releasing chlorofluorocarbon sprays we deplete the atmosphere's ozone and let in lethal quantities of ultraviolet radiation. By burning fossil fuels and felling forests we might warm the atmosphere, shifting climatic zones and spreading deserts. But in a few thousand years, variations in the Earth's tilt and distance from the Sun will cool the atmosphere, enlarging ice sheets and lowering the oceans. Later, continental drift will let warm ocean currents into polar regions. Then melting ice sheets may drown low land but cause a rebound of ice-laden Greenland and Antarctica.

Meanwhile, sea-floor spreading and subduction will see oceans grow and shrink. In 30 million years the Atlantic will be much wider while the Pacific will have narrowed.

Dramatic changes will affect the land. Already human overuse of soil accelerates erosion and the accumulation of sediments in offshore waters. Longer term, mountains will be worn to stubs, while new ones rise as continents collide.

Sun, Earth, and Ice
Changes in Earth's tilt and seasonal nearness to the Sun arguably produce ice ages.
A Earth far from the Sun in northern summer with northern hemisphere tilted slightly sunward – northern ice sheets grow.
B Earth close to the Sun in northern summer with northern hemisphere tilted far sunward – northern ice sheets melt.
a Earth in winter
b Earth in summer
c Sun

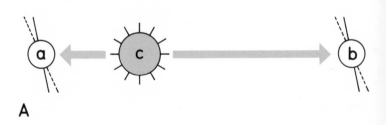

A

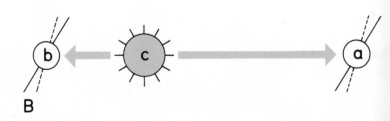

B

As Africa heads north, Europe's Alps may grow afresh, and the Rhine fault might split the continent. Iberia could be shoved into the Atlantic. The west Mediterranean may become a landlocked lake while squeezing destroys its east end and plants a mountain range from southeast Italy to Syria. Volcanic islands could appear offshore from Portugal to Norway.

Elsewhere, the Himalayas are still growing. In 15 million years, Australia might override Indonesia. In 25 million years, a north-south sea might split Africa in two. In 50 million years Los Angeles could be rafted north to join Alaska.

Billions of years will bring much more dramatic changes. A thinned asthenosphere may give the Earth a rigid crust. Because our planet's spin is slowing down each Earth day could be more than 50 of today's days long. The Moon will move away and hover over the same part of the Earth's surface, producing a fixed high tide. Eventually the Sun will swell up and engulf the Earth. Later still, our solar system may fall into a dense black hole in the middle of our galaxy.

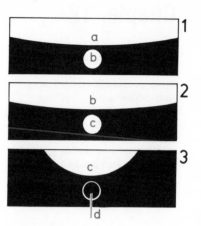

Black hole
Above: Diagrams stress the small size of a black hole.
1 Our Sun (**a**) compared to a white dwarf star (**b**)
2 White dwarf (**b**) compared to a neutron star (**c**)
3 Neutron star (**c**) compared to a black hole (**d**)
Below: A tiny but massive black hole (**a**) may distort the space-time fabric (**b**), creating an intense gravitational field that locks in electromagnetic radiation (**c**) and sucks in stars (**d**).

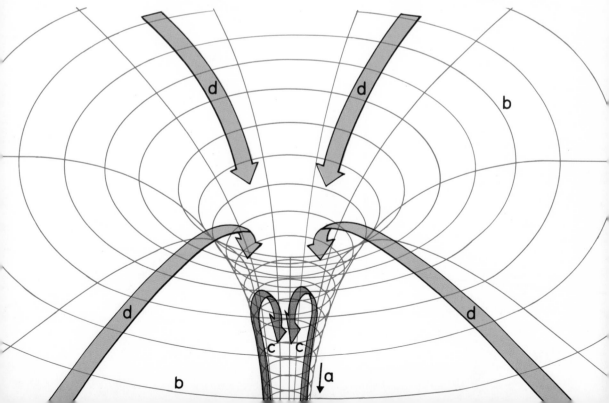

Chapter 12

ROCKS AND MAN

We briefly cover mapping
rocks, collecting minerals,
finding and extracting ores and
other useful substances, and
know-how used in putting
rocks to work. The book ends
with a short survey of pioneers
who helped us read the rocks,
and a brief worldwide guide to
museums and landmarks where
geology is on display.

Two 16th century woodcuts
showing mining activities.
(First published in *De Re
Metallica* by Georgius Agricola,
1556)

217

Mapping rocks

Mapping rocks involves detective work, for most solid rock (the "solid" geology) lies hidden under superficial deposits: residual remains of weathered rocks and "drift" laid down by rivers, glaciers, or wind. But solid rocks do show up here and there – in sea cliffs, river beds, quarries, road and railroad cuttings, boreholes, wells, pipeline trenches, and miners' spoil heaps. Even plowed fields may hold stones and boulders from the underlying rock. Elsewhere, the distribution of hills and valleys, springs, and even plants may indirectly hint at variations in the rocks below. Between them, such clues can indicate resistant and easily eroded rocks, faults, and layers of impermeable clay, or beds of limestone or sandstone.

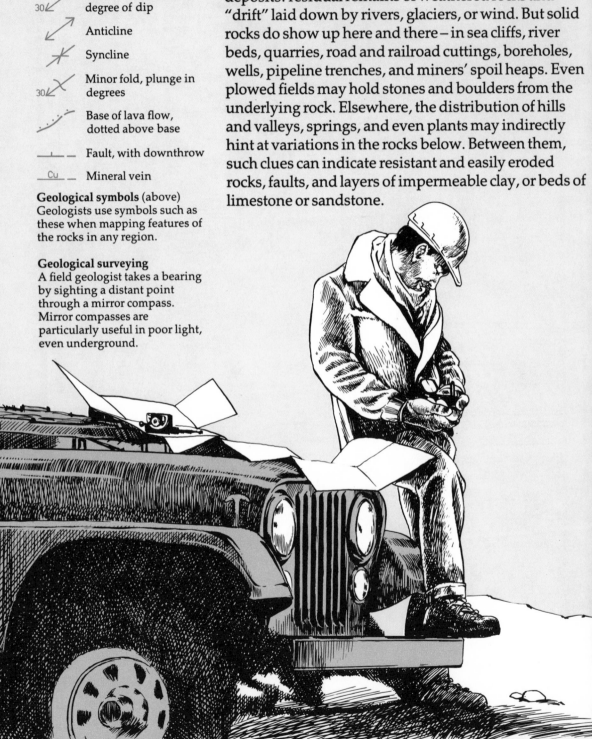

+   Horizontal strata

30   Inclined strata with degree of dip

  Anticline

  Syncline

30   Minor fold, plunge in degrees

  Base of lava flow, dotted above base

  Fault, with downthrow

Cu   Mineral vein

Geological symbols (above)
Geologists use symbols such as these when mapping features of the rocks in any region.

Geological surveying
A field geologist takes a bearing by sighting a distant point through a mirror compass. Mirror compasses are particularly useful in poor light, even underground.

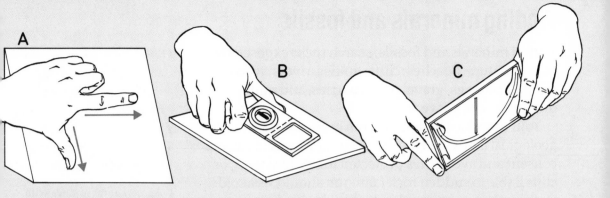

The field geologist visits rock exposures. He or she goes armed with at least a large-scale topographical map, compass, clinometer (or both combined in one), and rock-collecting tools described on pages 222–223. Between exposures, a hoe or hand auger may reveal rocks just below the soil. The geologist collects, identifies, and records rock and fossil samples, and plots exposures by compass on a map sheet. A clinometer will show dip (angle of tilt) of rock beds and plunge (tilt) of fold axes. A compass indicates strike (a horizontal compass direction at right angles to direction of dip). The mapmaker records angles of rock beds, faults, folds, joints, cleavage, and foliation, and notes features such as overturned beds, unconformities, intruded and extruded igneous rocks, and rocks metamorphosed by intrusions or other features.

Sometimes aerial photography aids observation in the field. Stereoscopic photographs stress features not always noticeable from the ground.

Once the geologist has recorded the types and angles of all exposed rocks he or she can then infer and map the hidden rocks between.

Strike and dip (above)
A The right-hand rule: when your thumb points down the dip, your index finger shows the bearing of the strike.
B Contact method for measuring strike: holding a compass horizontally, aligned parallel to strike, with compass edge against the rock surface or a map case laid on that to make it smooth and even.
C Using a clinometer to measure dip, at right angles to strike.

©DIAGRAM

Finding minerals and fossils

To find minerals and fossils, search rocks exposed by man or nature. Beaches, cliffs, gullies, rivers, road and railroad cuttings, gravel pits, quarries, and mine dumps may all prove fruitful.

Your first clues will come from guide books, geological maps, and local museums. Ask landowners' consent and avoid sites protected for their rarities, or cliffs liable to sudden rock falls. You should wear old clothes, strong shoes or boots, goggles, and perhaps protective helmet.

Walk slowly, gazing on the ground. "Rock hounds" will seek freshly broken rocks, rock cavities, veins of calcite, and weathered fragments of attractive minerals. Fossil hunters search for exposed rocks containing bits of fossil plant or animal. Some rocks teem with fossils, others hold a few tell-tale shiny or discolored shapes, yet other rocks are sterile.

All major types of rock offer hope for people seeking decorative minerals. For instance, among igneous rocks, granites hold well-shaped crystals, and lava flows may include gas cavities rimmed with gleaming calcite or agate. Among sedimentary rocks, limestones and some shales, hold geodes – lumpy concretions often with crystal-lined internal cavities. Metamorphic rocks may feature colorful serpentine and marble. Many kinds of rocks hold veins of minerals. Even pebbles on a beach might be worth examining for agate, amethyst, and onyx.

Only sedimentary rocks hold fossils. The richest hunting grounds are marine limestones, shales, and certain sandstones. Likely finds are bits of fossil sea shell, coral, and hard parts of other animals without a backbone. But fossil land plants and animals crop up in certain rocks – the waste from coal mines for example.

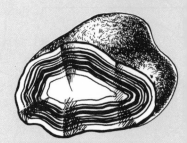

1 Agate, a form of silica with parallel colored bands – often found in volcanic rocks

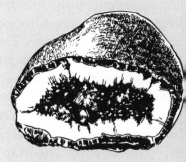

4 Geode broken open to reveal decorative crystals that rim an internal cavity

Finding minerals (right)
An amateur "rock hound" wears a hard hat to hunt for geodes among rock debris fallen from a cliff containing shale and limestone layers. Only splitting open lumps will show what actually lies inside.

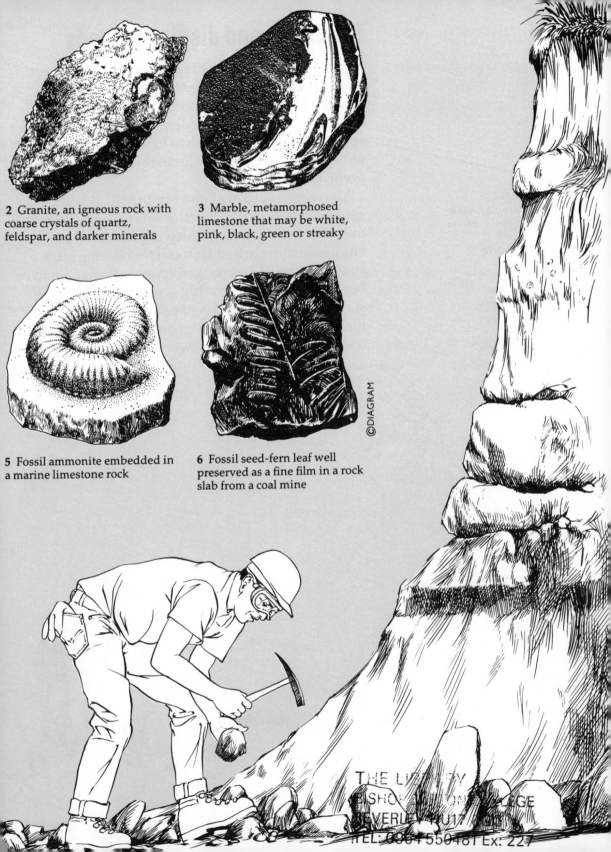

2 Granite, an igneous rock with coarse crystals of quartz, feldspar, and darker minerals

3 Marble, metamorphosed limestone that may be white, pink, black, green or streaky

5 Fossil ammonite embedded in a marine limestone rock

6 Fossil seed-fern leaf well preserved as a fine film in a rock slab from a coal mine

©DIAGRAM

Extracting and displaying finds

Finding a mineral or fossil specimen is just the start. Next you must free the specimen from its matrix (surrounding rock), transport it home, then clean, identify, and arrange it for display. (What follows applies more to minerals than fossils which are dealt with in more detail in our companion book on prehistoric life.)

You can just pick up loose bits of rock. But use a geological hammer, a chisel, or a punch and awl to free mineral specimens embedded in a matrix. A pocket knife may come in handy, too. Try choosing pure specimens with undamaged crystals. Items should be big enough – but not too big – to put in a display. A shovel, sieve, rake, and gold pan may help you wash for heavy minerals in stream beds. A hand lens is useful for examining small specimens.

Rock hound's toolkit
1 Geological hammer
2 Chisel
3 Pocket knife
4 Gold pan
5 Hand lens
6 Cloth bag
7 Plastic bags
8 Notebook
9 Knapsack
10 Plastic goggles

Arranging a collection
Some collectors put each specimen in a box (**A**) labeled to identify and date its contents. (Also sticking a label on each specimen prevents confusion should its box be lost.) Numbered boxes may be placed in rows in drawers of a cabinet (**B**).

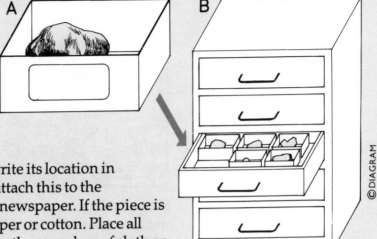

©DIAGRAM

After collecting any item write its location in ballpoint on sticky tape and attach this to the specimen. Then wrap this in newspaper. If the piece is fragile, first wrap in tissue paper or cotton. Place all specimens from one locality in the same bag of cloth or polythene, and label it.

Carry a notebook and pencil to record localities and mark these on a large-scale map.

Back home, wash dirt and stains from specimens. Identifying most means choosing from 200 common compounds. With experience you can whittle down the possibilities by noting features such as color, hardness, luster, cleavage, and perhaps fluorescence or specific gravity. (Field guide books, museum visits, and collecting trips with a museum class or rock and mineral club will all improve your diagnostic skills.)

Arrange your mineral collection according to a numbering system. Paint a small white square on each item and write its number on the square in Indian ink. Then place items in an ordered display on shelves or in shallow drawers. Keep minerals dust free to secure the best effect.

Hardness scale (right)
Softest (**1**) and hardest (**10**) minerals appear with others and everyday objects on a hardness scale devised by Austrian Friedrich Mohs in 1822. Each item scratches all those softer than itself.

#	Mineral	Object	Hardness
1	Talc		
2	Gypsum		
3	Calcite	Fingernail	2½
4	Fluorite	Silver	2½-3
5	Apatite	Teeth	5
6	Orthoclase	Penknife	5½
7	Quartz	Glass	6
8	Topaz		
9	Corundum		
10	Diamond		

Useful minerals

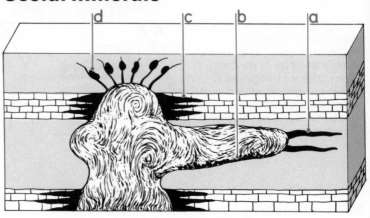

d c b a

Ores and magma (right)
A magma body injected into
country rocks produces ores in
several ways.
a Dense minerals settle in
magma as it cools.
b Slow-cooling minerals
crystallize in fissures.
c Magma minerals replace
country rock in metamorphism
d Hydrothermal deposits of
magma minerals form in
fissures.

Some field geologists seek ores – rocks rich enough in
metals or certain other elements to be worth mining
and separating from gangue (unwanted substances).
Useful nonmetallic rocks and minerals include beds of
gypsum, salt, and limestone. Igneous, metamorphic
and sedimentary rocks all hold valuable substances,
but only special areas reward a search.

Many ores form in or near a mass of molten magma,
so igneous and nearby rocks are often fruitful hunting
grounds. In slow-cooling basic magma, among the first
minerals to crystallize and settle are apatite, magnetite,
and chromite (respective sources of phosphorus, iron,
and chromium). These ores occur as bands in sills and
dikes.

Ores form, too, where a molten granite batholith
injects hot, mineral-rich fluids under pressure into
nearby rocks. As the fluids cool, their crystallizing
minerals fill cracks, producing veins or groups of veins
called lodes. Hot mineral-rich solutions losing heat
and pressure as they filter out of granite replace nearby
rocks with a sequence of tin, copper, lead, zinc, iron,
gold, and mercury.

Certain useful substances occur as precipitates or
evaporites. Thus hot springs precipitate the mercury
ore cinnabar, and precipitation creates manganese
nodules on the sea bed. Evaporation of seas and inland
lakes builds thick beds of gypsum, anhydrite, halite
(rock salt) and potash.

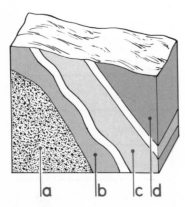

a b c d

Ore sequence
A block diagram shows the
mineralized zones produced as
temperature and pressure
changes affect mineral-rich
solutions moving out from
molten granite.
a Granite mass
b Tin deposits
c Copper deposits
d Lead and zinc deposits

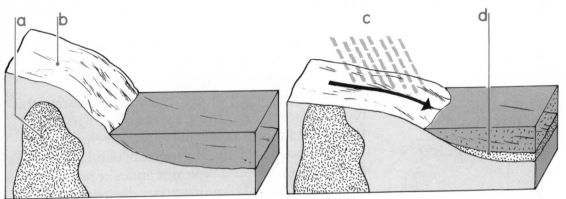

Then there are residues or sediments. In the tropics, weathering breaks down aluminum silicate rocks, yielding bauxite, the raw material for aluminum. Weathering also forms some ores of iron and manganese. Percolating groundwater can deposit rich copper ores. Rivers and coastal waves sort particles of heavy minerals, dumping them on stream beds and beaches. Such placer deposits provide much of the world's tin, also some gold and platinum, with diamonds and other gemstones.

Sedimentary ores (above)
1 Iron-rich pluton (**a**) inside a mountain (**b**)
2 Weathering and erosion (**c**) expose the pluton and wash dissolved iron into the sea. Precipitated iron oxide (**d**) accumulates in the sea bed.

Mineral reserves (below)
This world map plots land distribution of important metals and nonmetals.

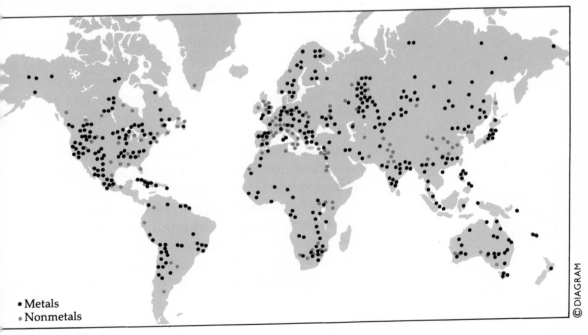

• Metals
• Nonmetals

© DIAGRAM

225

Gemstones

Gemstones are minerals prized for beauty, durability, and rarity. Their worth depends on scarcity, color, purity, brilliance, hardness, and demand. Most stones are cut and polished so they glow and sparkle, then set in gold or otherwise as articles of jewelry.

Gemstones largely form in igneous or metamorphic rock from elements combining as they cool in gas pockets or superheated water. The coarse-grained igneous rock pegmatite is a source of beryl, tourmaline, and topaz. Volcanic andesite holds cavities where opal grew. Volcanic activity deep down produced the intense heat and pressure that forged the diamonds now found in pipes of kimberlite, a version of the rock peridotite. Emeralds are a transparent type of beryl that occurs in mica-schist, a metamorphic rock.

Jewelers identify gems by color, shape, hardness, refractive index ("light-bending power"), and specific gravity. Diamond, ruby, emerald, and sapphire – all transparent gemstones – rank with opaque opal as the precious stones. Semiprecious stones include agate, amethyst, garnet, and peridot. Many gemstones are simply spectacular forms of ordinary-looking minerals. For instance, diamond is a pure, hard form of carbon – the substance coal is made of, and ruby and sapphire are just transparent, colored forms of corundum, much of which is drab or colorless. Metal oxides tint most gemstones, but diamonds get their color from a defect in their crystal structure.

Craftsmen cut and polish gems to bring out their special features. Translucent and opaque stones may get the rounded shape called cabochon. Transparent gems are faceted so that they reflect and bend the light. Each expert cuts a stone by grinding with a hard abrasive; diamond dust alone is hard enough to cut a diamond. Most stones are cut in one of four main styles, called brilliant, step, mixed (brilliant and step), and rose.

Major styles of cut
(Side views inset)
1 Brilliant
2 Step cut
3 Rose cut
4 Cabochon

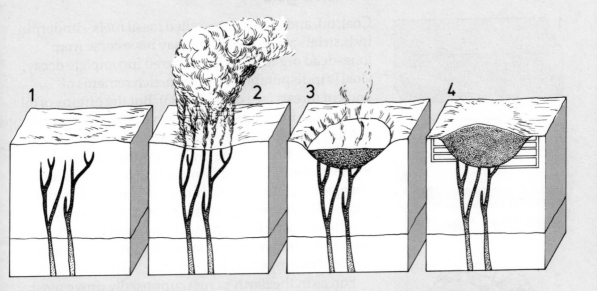

Diamonds' origins (above)
1 Intense heat and pressure create diamonds deep down in kimberlite pipes.
2 Gas exploding in fissures leaves a hollow in the land.

3 Kimberlite containing diamonds wells up and fills the hollow.
4 Miners sink shafts to reach the lower levels of the kimberlite.

The world's gems (below)
This map marks major sources of the five precious stones. Diamonds largely occur in old cratons but some were washed out as placer deposits.

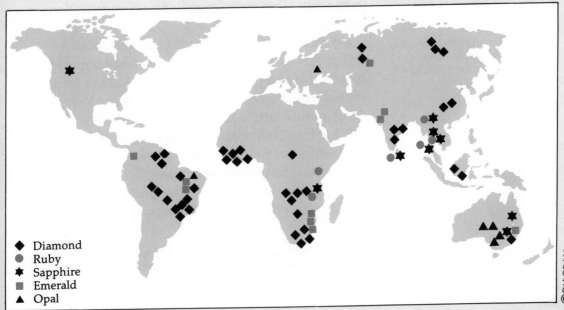

◆ Diamond
● Ruby
★ Sapphire
■ Emerald
▲ Opal

©DIAGRAM

Oil and gas

Coal, oil, and gas – the so-called fossil fuels – underpin industrial society today. All may have come from long-dead organisms that suffered incomplete decay. Coal is indisputably the carbon-rich remains of ancient forests (see pages 82–83). But the origins of oil and gas are more obscure and have been open to dispute.

Most geologists believe that natural gas and oil derive from tiny marine organisms that died and sank to the sea bed many millions years ago. Compaction changed surrounding sediments to mudstones and shales. The resulting heat and pressure probably produced bacterial processes that helped transform the organisms into hydrocarbons – compounds mainly made of hydrogen and carbon.

Forces in the Earth's crust supposedly drove most hydrocarbons from the rock in which they formed. They percolated up through permeable sand, sandstone, or limestone until trapped below a layer of impermeable rock. So the permeable rock below the trap became a fossil fuel reservoir. Here, natural gas floats on a layer of sticky to runny black to yellow liquid – a complex mix of hydrocarbons that we call petroleum. This petroleum floats on an even denser substance, water.

University astrophysicist Thomas Gold rejects the "squashed fish" theory. Gold argues that natural gas and oil originated in the Earth's formation. He believes that enough of both fuels lies locked up deep in the Earth to last us millions of years. In the late 1980s, a Swedish deep-drilling project with Gold as an adviser found gas traces in cavities 3.7mi (6km) below the surface. But more proof was needed to convince the skeptics. Most geologists still expected recoverable oil and gas supplies to dwindle sharply in a few decades.

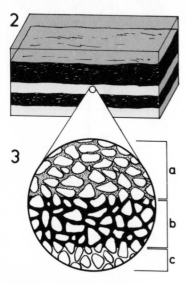

Hydrocarbons forming (above)
1 Dead organisms sink to the seabed.
2 Rocks cover them.
3 Bacterial action produces gas (a) and oil (b) above water (c) between sandstone particles (shown enlarged).

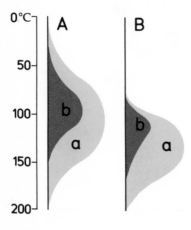

Amounts and temperatures
Tests show relative amounts of gas (a) and oil (b) produced as temperatures increase with depth of burial.

A Amounts from marine plants
B Amounts from land plants

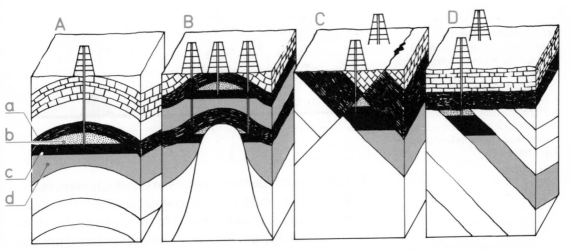

Reservoir rocks (above)
Here are wells drilled down through impermeable rock (**a**) and gas (**b**) to oil (**c**) trapped over water (**d**) in four types of situation.

A Oil in anticline
B Oil in rocks pushed up by a salt dome
C Oil trapped by fault
D Oil in tilted rocks trapped by unconformity

The world's oilfields (below)
Most of the oil-producing areas of the world lie in sedimentary basins. Oil does not survive igneous or metamorphic activity.

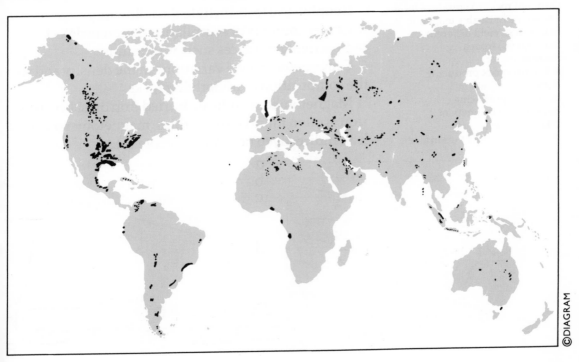

©DIAGRAM

Geological prospecting

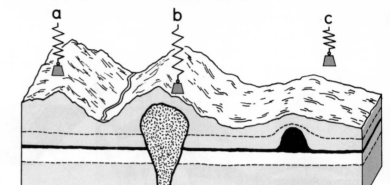

Gravimetric readings
A weight hangs from a coiled spring whose length varies with the force of gravity exerted by the rocks beneath.
a Normal reading
b High reading from dense igneous rock near the surface
c Low reading from low density salt dome

To track down useful minerals or other substances below the ground the geologist gathers all information already known about the area to be explored. Next, he or she maps this geologically, noting surface clues like faults and gangue minerals including quartz. Then the geologist can bring to bear any of a battery of tests.

Geochemical tests analyze rock and other samples for trace elements that may lead the geologist to a major ore body.

Geophysical tests include the following.

Geiger counters or scintillation counters detect radioactive substances such as uranium.

Gravimeters reveal variations in the density and so the composition of underlying rocks.

Magnetometers indicate buried iron ores. Because iron is often found with sulfides, magnetometers may lead indirectly to non-ferrous metals, too.

Magnetometer readings (below)
Different rocks (**A**) produce local variations in Earth's magnetic field and so yield different readings (**B**) from a magnetometer.
a Country rock producing regional magnetism
b Topsoil producing background magnetism
c Deeply buried ores
d Ores just below the surface

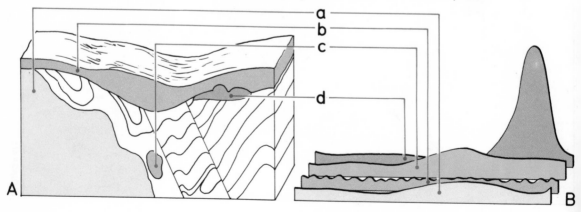

Electrical surveys show certain ores affecting natural ground currents related to the Earth's magnetic field.

Seismic surveys test for various deposits, including oil, gas, and coal. Seismic surveying involves setting off explosions or vibrations that send shock waves down into the ground and timing their return from surfaces that bend or bounce them back. The speed of their return indicates the depth and nature of the rocks below.

All these prospecting methods merely hint at what lies underground. Only exploration can prove an ore is actually there and rich and big enough to be worth mining. If prospecting gives encouraging results, exploration follows. This means drilling sample cores or digging trial trenches to find out if development would pay. The next pages show just what development is likely to involve.

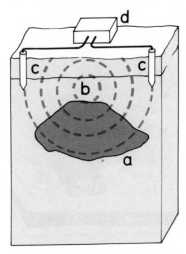

Electrical survey (above)
An ore deposit (**a**) affects natural ground currents (**b**) flowing between buried electrodes (**c**). A millivoltmeter (**d**) registers voltages at the electrodes.

Seismic survey (left)
Dynamite exploded underground (**a**) produces shock waves, bent and bounced back by rock layers (**b,c**) at different depths to surface pickups (**d**). These relay the data to a recording truck (**e**).

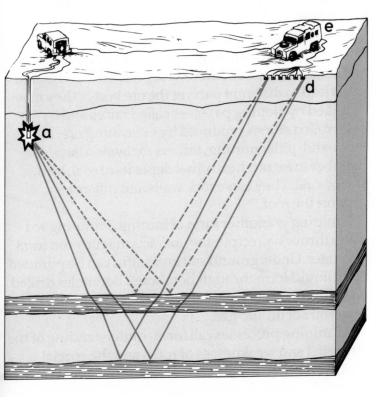

Mining

Some ores and useful nonmetallic minerals and rocks can be collected from the surface. Mining others may mean burrowing deep into the Earth's crust.

Surface mining includes placer mining, dredging, strip mining, quarrying, and open-pit mining. Placer mining uses running water to sort heavy minerals such as tin, gold, and platinum from sand and gravel. Dredging employs mechanical scoops or buckets to raise sand or gravel from a pond or lake, before washing extracts heavy-mineral deposits. Strip mining strips off surface material so that power shovels can remove the underlying ore or coal. Quarrying uses saws, wedges, or explosives to free blocks of building stone from deposits near the surface, and power shovels to scoop up sand or gravel. Open-pit mining extracts ores from hard rock by cutting benches (steplike ledges) in a mountain.

Underground mining involves boring and blasting a hole into the Earth's crust to reach an ore body. A nearly horizontal hole is called an adit, a vertical hole is a shaft. Horizontal passages following a vein are drifts. In level-and-shaft mining, a deep shaft leads to horizontal tunnels at different depths. These so-called levels lead to different parts of the ore body. They may be linked by sloping passages called raises and by holes called stopes produced by excavating ore. In room-and-pillar mining, miners excavate a large chamber to extract horizontal deposits of coal, lead, salt, or zinc. They leave rock walls and pillars to support the roof.

Pumping is another form of mining. Pumping sea water through precipitators extracts magnesium from sea water. Underground salt and sulfur can be pumped up in liquid form through boreholes. Boreholes drilled on land and on continental shelves also tap rock reservoirs of oil and gas.

All mining processes call for an understanding of the strengths and weaknesses of rocks and the special problems posed by faults and folds.

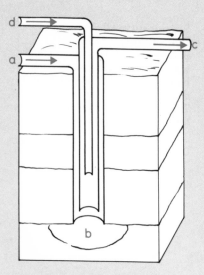

Strip mining (above)
A power shovel (**a**) strips away superficial soil and rock called overburden (**b**) to expose a coal seam (**c**) just below the surface.

Pumping up sulfur
Hot water flowing down the outer pipe (**a**) of three concentric pipes melts underground sulfur (**b**) which rises through a middle pipe (**c**), forced up by compressed air from the inner pipe (**d**).

Underground mine (right)
Most items depict those named in the text.
a Adit b Shaft
c Levels d Ore body
e Stope f Raise
g Winze (driven down from a level) h Drift

Room and pillar mining
Below: This view stresses coal pillars left by first workings, but omits most of the rock roof they support.
a Unworked coal seam
b Coal pillars
c Rooms left by excavated coal
d Rock roof
e Coal conveyor
f Coal stockpile

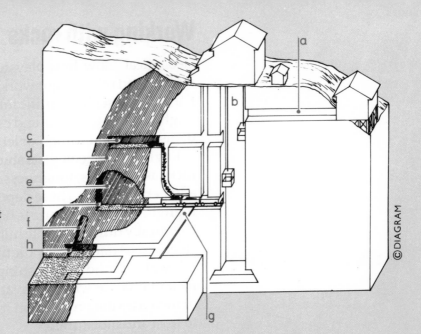

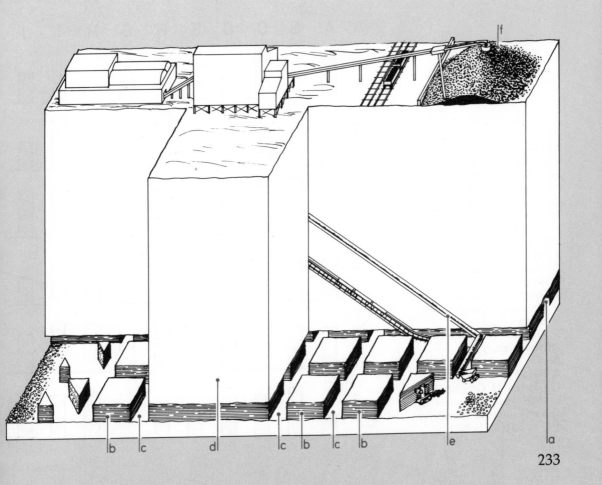

©DIAGRAM

233

Working with rocks

Geological know-how plays a key role in modern mining and other major works involving excavation. For without such scientific information engineers might make dangerous and costly errors that burst dams, topple buildings, or let tunnel roofs collapse.

Engineering geologists study such properties of rocks as hardness, toughness, elasticity, durability, and permeability. These depend largely on mineral ingredients. Thus quartz is harder than steel and chemically durable, while mica is much softer than steel and its dark variety biotite is readily attacked by rainwater. Grinding, crushing, and other tests hint at a rock's behavior when excavation or building reduces or increases the pressure on it, and perhaps exposes it to frost and moisture.

Properties of rocks
Bar heights indicate actual (1-2) and relative (3-6) values for 10 types of rock used in roads and concrete. In 1-4 the higher the bar the higher the quality; in 5-6 the reverse.
Properties:
1 Specific gravity
2 Crushing strength
3 Hardness (abrasion)
4 Toughness (impact)
5 Abrasion value-dry (dry attrition value)
6 Abrasion value-wet (wet attrition value)
Rocks:
A Basalt B Flint
C Gabbro D Granite
E Gritstone F Hornfels
G Limestone H Porphyry
I Quartzite J Schist

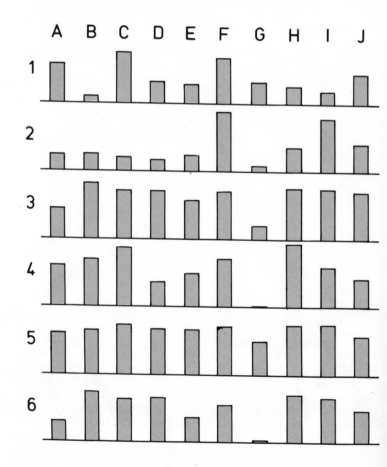

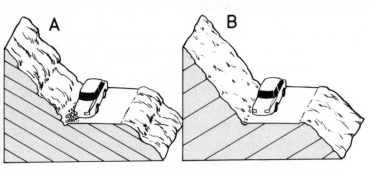

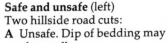

Safe and unsafe (left)
Two hillside road cuts:
A Unsafe. Dip of bedding may produce collapse.
B Safe. Dip of bedding protects against collapse.

Similarly soil mechanics deals with properties of clays, silts, sands, gravels, organic soils, and peat – their capacities to carry loads, and liabilities to settlement, compaction, permeability, and frost.

Then, too, geological mapping of a chosen area reveals local weaknesses like faults, joints, bedding planes, landslides, and soil-filled river channels snaking through a bed of solid rock.

All this information helps to show where engineers can safely carve out tunnels, cuttings, quarries, and canals, or excavate foundations for dams, bridges, and large buildings. Geological studies also show sites best avoided or requiring special structures. Thus engineers can throw a slim, concrete arch dam across a narrow gorge walled by strong, elastic rock. But such a dam would burst if built on soft, weak shale. Here, they would raise a massive earth- and rock-filled dam instead. Similarly, New York's resistant metamorphic rock provides a firmer base for skyscrapers than the soft clay London stands on.

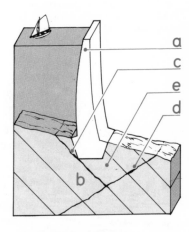

A dam that burst (above)
a Concrete arch dam
b Bedrock
c Crack along foliation plane
d Crushed seam
e Rock wedge
Water seeping under pressure into (**c**) dislodged the wedge and burst the dam.

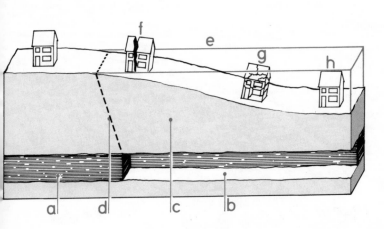

Mining subsidence (left)
This diagram shows three possible effects on buildings.
a Unworked coal seam
b Worked coal seam
c Subsiding rock
d Limit of subsidence
e Presubsidence land level
f Tension damage
g Compression damage
h Vertical displacement

©DIAGRAM

Man-made rocks

Rocks and the landforms made by their erosion helped decide where people built towns, ports, and cities. Now, the environmental geologist works with rocks to find supplies of water and building materials, and to check building-site stability, assess local risks of flood or earthquake, trace the origins of toxic chemicals and polluted water, and choose safe dumps for refuse burial.

The immense extent of modern buildings, airports, roads, and railroads means that these in turn affect the surface of the continents. Across mighty tracts of land,

Man-made landscape
City skyscrapers, streets, and other structures are made from rocks extracted from the ground and reconstructed into concrete, steel, glass, bricks and other man-made substances.

workers have masked the native soil with man-made rocks. Their chief materials are clay bricks and tiles; walls and floors of concrete (sand, stones, lime and other substances bound by water); roads of stones set in bitumen; glass windows of silica, lime, and soda; and steel rods and girders of iron with added elements.

Most construction works affect the natural processes that wear away and build dry land.

Some projects curb erosion: dams raise the base to which rivers erode their beds; groynes and sea walls hamper the destructive work of waves. By drying swamps and marshes, drainage actually speeds up land formation. Reversing the erosion process altogether, offshore dredging scoops building sand and gravel from the continental shelf, and Dutch engineers have won much of Holland from the sea.

But human handiwork accelerates erosion, too – by mining and quarrying, and through farming malpractice that loosens soil until rain or wind wash or blow the soil away. Also, miners extract fossil fuels and certain minerals far faster than the Earth replaces them.

For ill or good, our species is a geologic agent strong enough to tamper with our planet's crust.

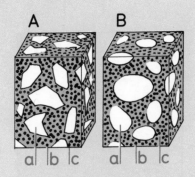

Man-made rock (above)
Concrete is a man-made equivalent of conglomerate.
A Concrete
B Conglomerate
a Clasts
b Matrix separating clasts
c Cement bonding clasts

Man-made land (below)
A Area of the Netherlands above sea level
B Actual land area today, increased by swamp drainage and reclamation from the sea. Tinted area shows land below sea level.

©DIAGRAM

237

Great geologists 1

These four pages list selected geologists and others who helped reveal the shape or structure of the Earth and worked out how its surface features formed. No list this short can be complete. Ours takes in mere dozens of the hundreds of North American and European rock detectives who have pieced together the puzzle of our planet's past.

From Agricola's noted work on mining *De re metallica*

Agassiz, Louis (1807–73) Swiss scientist who argued that ice sheets had covered much of Pleistocene North America and northern Europe.
Agricola, Georgius (1494–1555) German father of mineralogy. He studied ores, recognizing mineral veins as deposits left by rising solutions.
Airy, Sir George Biddell (1801–92) British astronomer who in the 1850s laid a basis for the theory of isostasy.
Alberti, Friedrich August von (1795–1878) German geologist who named the Triassic System in 1824, from a tripartite division of rocks.
Arduino, Giovanni (1714–95) Father of Italian geology who coined the name "Tertiary," later given to a rock system and geological period.
Aristotle (384–332 BC) Greek philosopher who believed the Earth was spherical because it cast a curved shadow on the Moon.
Barrell, Joseph (1869–1919) American geologist who declared that much sedimentary rock did not form under oceans, and molten magma interacting with the crust formed fissures and new surface features. He coined the terms "lithosphere" and "asthenosphere" and was the first geologist fully to realize the potential of radioactive dating.
Barrow, George Nineteenth-century Scottish geologist who found in the 1880s that different temperatures produced different metamorphic rocks from the same ingredients.
Beaumont, Élie de (1798–1874) French geologist who began a geological map of France and mistakenly proposed that mountain ranges were forced up by compression caused by shrinkage of the Earth.
Benioff, Hugo (1899–1968) American seismologist whose studies revealed the Benioff zone where earthquake foci descending from an ocean trench into the mantle mark subduction of an oceanic plate.
Bertrand, Marcel-Alexandre (1847–1907) French geologist who showed that formation of such mountain ranges as the Alps involved massive folding of the Earth's crust.
Beyrich, Heinrich Ernst (1815–96) German paleontologist who in 1854 introduced the term Oligocene.
Bowen, Norman Levi (1887–1956) American experimental petrologist who showed the order in which minerals crystallize from basaltic magmas as these cool, so producing different igneous rocks.
Brongniart, Alexandre (1770–1847) French geologist who in 1829 coined the name "Jurassic ground" for rocks later put in the Jurassic System.
Bryan, Kirk (1888–1950) American geologist and geomorphologist who studied arid-climate landform evolution, and with South African geomorphologist Lester King pioneered the theory of parallel slope retreat often credited to Austria's Walther Penck.
Buch, Leopold von (1774–1853) German geologist who produced a geological map of Germany in 1824. He studied volcanoes and coined the term "andesite" for Andean volcanic rock of the type now known to be produced at subduction zones.

Alexandre Brongniart

Leopold von Buch

Carey, Samuel William Twentieth-century Australian geologist who proposed the controversial theory that the Earth has been expanding.

Conybeare, William Daniel (1787–1857) British geologist who with William Phillips in 1822 called certain strata "Carboniferous."

Copernicus, Nicolaus (1473–1543) Polish astronomer who argued that the Earth was a rotating planet orbiting the Sun.

Cushman, Joseph (1881–1949) American paleontologist who began using foraminiferans for relative rock dating.

Cuvier, Georges (1769–1832) French anatomist and paleontologist whose discovery that many fossil invertebrates were now extinct aided the correlation and relative dating of sedimentary rocks.

Dana, James Dwight (1813–95) American scientist who classified minerals, coined the term "geosyncline," studied coral-rock formation, and theorized about the evolution of the Earth's crust.

Darwin, Charles Robert (1809–82) British naturalist who developed the theory of biological evolution by natural selection. He rightly judged that seabed subsidence created coral atolls, and thought repeated earthquakes could build mountain ranges.

Davis, William Morris (1850–1934) American geographer and geologist; a founder of geomorphology (scientific landform studies) who stressed the so-called cycle of erosion.

Descartes, René (1595–1650) French scientist-philosopher who thought the Earth possessed a layered structure and had once been molten.

Deshayes, Gérard Paul (1797–1875) French conchologist whose work on fossil shells helped subdivide the Tertiary Period into epochs.

Desnoyers, Jules French geologist who in 1829 separated Quaternary from Tertiary rocks.

Dewey, John F. British scientist who with the American John M. Bird in 1970 showed how plate tectonics had created mountain belts.

Dietz, Robert Sinclair Twentieth-century American oceanographer who helped pioneer and named the theory of sea-floor spreading.

Du Toit, James Alexander Logie (1878–1948) South African geologist who suggested that Pangea had split into (northern) Laurasia and (southern) Gondwana, separated by the Tethys seaway.

Dutton, Clarence Edward (1841–1912) American geologist, seismologist, and vulcanologist who advanced and named the theory of isostasy.

Eratosthenes (276–194 BC) Greek astronomer who measured the Earth's size with remarkable accuracy.

Eskola, Pentti Eelis (1883–1964) Finnish petrologist who proposed the idea of metamorphic facies – rocks of different composition but metamorphosed under similar conditions.

Geer, Gerard Jakob, Baron de (1858–1943) Swedish geologist who pioneered varve-counting to date sediments in glacial lakes.

Gilbert, Grove Karl (1843–1918) An American founder of landform studies (geomorphology) who coined the term "orogeny" for mountain building.

Gressley, Amanz (1814–65) Swiss geologist who coined the term "facies" for lateral variations seen in rocks of the same age.

Guettard, Jean Étienne (1715–86) French mineralogist who produced arguably the first geologic maps.

Hess, Harry Hammond (1906–69) American geologist and geophysicist who in the 1940s discovered guyots (flat-topped submarine peaks) and in 1960 published arguments for sea-floor spreading.

Samuel Carey

Clarence Dutton

Harry Hess

©DIAGRAM

239

Great geologists 2

James Hutton

Sir Charles Lyell

Alcide d'Orbigny

Holmes, Arthur (1890–1965) British geologist and geophysicist who put dates to the geological time scale as early as 1913. In 1928 he argued that convection currents in the mantle moved continental crustal blocks which collided, forming mountains and leaving crustal gaps plugged by new rock produced in ocean basins.

Hutton, James (1726–97) Scottish geologist who pioneered uniformitarianism – belief that forces still at work had caused geological change over a vast span of time. He contributed to the understanding of how igneous rocks are formed.

Lapworth, Charles (1842–1920) British geologist who in 1873 identified the Ordovician System by means of fossil graptolites.

Lemaître, Georges (1894–1966) Belgian astronomer and cosmologist who in 1927 proposed a Big Bang theory for the origin of the universe.

Le Pichon, Xavier French marine geophysicist who with W.J. Morgan in 1968 formulated the plate tectonics hypothesis, involving rigid plates that moved.

Lyell, Sir Charles (1797–1875) British geologist who in the 1830s introduced the terms Eocene, Miocene, Pliocene, Pleistocene, and Holocene (Recent). Modern geology owes much to his *Principles of Geology.*

McKenzie, Dan Twentieth-century British geophysicist who with R. L. Parker in 1967 related sea-floor spreading, transform faults, and island arcs to (lithospheric) plates with interacting boundaries.

Mohorovičić, Andrija (1857–1936) Croatian geophysicist whose earthquake studies led to the discovery of the Mohorovičić Discontinuity, a boundary between the crust and mantle.

Morgan, William Jason American geophysicist, coauthor with Le Pichon, arguing in 1968 that "rigid blocks" (lithospheric plates) slid over a weak asthenosphere which acted as a lubricant. In 1971 he proposed that the Hawaiian Islands grew from one of many fixed "mantle plumes" or hot spots underlying moving plates.

Murchison, Sir Roderick Impey (1792–1871) British geologist who recognized the Silurian and Permian systems.

Newton, Sir Isaac (1643–1727) English scientist and mathematician whose theories of motion and gravitation helped explain the Earth's elliptical path, worked out by Johannes Kepler (1571–1630).

Omalius d'Halloy, Jean-Baptiste-Julien (1783–1875) Belgian geologist who produced systematic subdivisions of geological formations and gave Cretaceous rocks that name.

Oppel, Albert (1831–65) German geologist and paleontologist who subdivided stages into zones.

Orbigny, Alcide Dessalines d' (1802–57) French founder of micropaleontology, who divided geological formations into stages.

Penck, Albrecht (1858–1945) German geographer and geologist who founded Pleistocene stratigraphy. He also helped pioneer and reputedly named geomorphology (landform studies).

Penck, Walther (1888–1923) Austrian geomorphologist who argued that straight, convex, and concave hillslopes reflected variations in the balance between rates of land uplift and denudation.

Phillips, John (1800–74) British geologist who in 1840 gave names to the Mesozoic and "Kainozoic" (Cenozoic) eras.

Post, Ernst Jakob Lennart von (1884–1951) Swedish geologist who with the botanist G. Lagerheim helped start the stratigraphic use of palynology (pollen studies).

Ray, John (1627–1705) English scientist who published an explanation of how springs occur.

Schimper, Wilhelm Philipp (1808–88) German paleontologist who in 1874 gave the Paleocene Epoch its name.

Schlotheim, Ernst von (1764–1832) German paleontologist, a pioneer in using fossils to find the relative ages of rock layers.

Sedgwick, Adam (1785–1873) British geologist who named the Paleozoic Era and the Cambrian and (with Murchison) Devonian systems.

Smith, William (1769–1839) So-called father of English geology. He used fossils to identify sedimentary rock layers, and produced the first geological map of England and Wales (1815).

Steno, Nicolaus (1638–86) Danish geologist who grasped that many rocks derived from sediments laid down in sequence in horizontal layers, and contained remains of ancient living organisms. He realized that running water was the main agent shaping landscapes.

Suess, Eduard (1831–1914) Austrian geologist who coined the terms "sima" and "sal" (later changed to "sial") for mantle and crust; located shields and orogenic belts around the world; and argued for northern and southern prehistoric supercontinents ("Atlantis" and "Gondwanaland") separated by a "Sea of Tethys."

Vinci, Leonardo da (1452–1519) Italian artist, engineer, and polymath who argued that marine fossils in upland rocks showed these must have formed below the sea.

Vine, Fred British geophysicist who with Drummond Matthews in 1963 argued that bands of normal and reverse magnetism in oceanic rocks showed sea floor formed at different times.

Wegener, Alfred Lothar (1880–1930) German meteorologist and geophysicist who in 1912 announced his theory later known as continental drift, involving splitting of a prehistoric supercontinent, Pangea, into modern continents.

Werner, Abraham Gottlob (1750–1817) German geologist who popularized Neptunism, a mistaken theory claiming that almost all rocks had been precipitated from the water of an early universal ocean.

Williams, Henry Shaler (1847–1918) American paleontologist who in 1891 introduced the term "Pennsylvanian" for Upper Carboniferous rocks in North America.

Wilson, John Tuzo Twentieth-century Canadian geophysicist who contributed to plate tectonics theory, and in the 1960s coined the geological terms "plates" and "transform faults."

Winchell, Alexander (1824–91) American geologist who in 1870 introduced the term "Mississippian" for Lower Carboniferous rocks in the Mississippi Valley.

Eduard Suess

Alfred Wegener

John Tuzo Wilson

©DIAGRAM

Geology displayed 1

Countless museums and some mines and quarries show or store rocks and minerals. Here six pages list a brief worldwide selection of collections. Items appear in alphabetical order of country and then city. Natural phenomena are grouped separately and marked by asterisks. Not all places are always open, so check access before you pay a visit.

Ayers Rock, Australia

ALGERIA
* **Ahaggar Mountains** Exposed part of the West African Shield
ANTARCTICA
* **Mt. Erebus** Antarctica's only active volcano
* **Lambert Glacier** The world's longest glacier
* **Ross Ice Shelf** The world's largest floating ice shelf
ARGENTINA
Buenos Aires: Museum of Mineralogy and Geology
* **Aconcagua** The Americas' highest peak, partly volcanic
AUSTRALIA
Ballarat, Victoria: Gold Museum
Coober Pedy, South Australia: Umoona Opal Mine Museum
Kalgoorlie, Western Australia: Hainault Tourist Gold Mine
Melbourne: Geological Museum; Museum of Victoria
Sydney: Geological and Mining Museum
* **Ayers Rock, Northern Territory** Vast sandstone monolith
* **Great Barrier Reef, Queensland** Largest living coral reef
AUSTRIA
Halstatt: Prehistoric Museum Local salt mining, etc
Innsbruck: Zeughaus Tirol Folk Museum
Klagenfurt: Mining Museum Local mining history
BELGIUM
Antwerp: Provincial Diamond Museum
Liège: Museum of Iron and Coal
Tournai: Museum of Paleontology and Prehistory
BOLIVIA
La Paz: Mineralogical Museum
BRAZIL
Rio de Janeiro: Museum of Geology and Mineralogy
* **Amazon River** The world's largest river, not all in Brazil
CANADA
Dawson City, Yukon: Dawson City Museum
Edmonton, Alberta: Museum of Geology
Knighton, Ontario: Geological Sciences Museum
Saskatoon, Saskatchewan: Geological Museum
Vancouver: M.Y. Williams Geological Museum
* **Niagara Falls, Ontario** Famous waterfall
CHINA
* **Shan-xi Loess Region** Thick windblown loess deposits
* **Yunnan Rock Forest** Tropical karst landscape
COLOMBIA
Bogotá: Museum of Minerals; National Geological Museum

Niagara Falls, Canada

Yunnan Rock Forest, China

DENMARK
Copenhagen: Geological Museum
ECUADOR
Quito: Petrographic Museum
EGYPT
Cairo: Geology Museum
FRANCE
Paris: Mineralogy Museum of the Paris School of Mines
* **Mont Blanc, Alps** Western Europe's highest peak
* **Cirque of Gavarnie, Pyrenees** Glaciated rock amphitheater
* **Fountain of Vaucluse, Provence** Underground source of the Sorgue
River – the original vauclusian spring
FRENCH POLYNESIA
* **Bora-Bora, Society Islands** Reef-fringed volcanic island
GERMANY (EAST)
Karl-Marx-Stadt: Underground Rock Cathedral Limestone quarry
GERMANY (WEST)
Aalen: Geological and Paleontological Museum
Bochum: German Mining Museum
Bonn: Mineralogical-Petrological Museum
**Munich: Bavarian State Collection for General and Applied Geology;
State Mineral Collection**
Münster: Geological-Paleontological Museum; Mineralogical Museum
Tübingen: Mineralogical Collection Exhibition
* **Rhine River** Part in a rift valley between block mountains
GHANA
Legon: Museum of the Department of Geology
GREECE
Athens: Museum of Mineralogy and Petrology
* **Thera (Santorini)** Island remnant of a vast exploded volcano
GUINEA
Conakry: Conakry Geological Museum
ICELAND
* **Strokkur** One of Iceland's most active geysers
* **Surtsey** Volcanic island that appeared in 1963
* **Vatnajökull** Europe's largest ice cap
INDIA
Lucknow: Geological Museum (two so named)
INDONESIA
Bandung: Geological Museum
* **Anak Krakatoa** Volcanic island on the site of Krakatoa which exploded
cataclysmically in 1883
IRELAND
Dublin: Trinity College Department of Geology
ISRAEL
* **Dead Sea, Israel/Jordan** Lowest and one of the saltiest bodies of water,
in the (Afro-Asian) Great Rift Valley
ITALY
Bologna: Museum of Mineralogy
**Catania: Geological Museum; Museum of Mineralogy and Petrography;
Museum of Vulcanology**
**Florence: Museum of Geology and Paleontology; Museum of
Mineralogy and Lithology**
Modena: Museum of the Institute of Mineralogy
Naples: Museum of Mineralogy and Zoology

Fountain of Vaucluse, France

Thera, Greece

Dead Sea, Israel/Jordan

©DIAGRAM

Geology displayed 2

Matterhorn, Italy/Switzerland

Fujiyama, Japan

Mt. Everest, Nepal

Parma: Museum of Mineralogy and Petrography
Pavia: Museum of Mineralogy and Petrography
Rome: Museum of the Institute of Mineralogy
Turin: Museum of Geology and Paleontology
* Mt. Etna Europe's highest active volcano
* Lardarello Hot Springs, Tuscany
* Matterhorn, Italy/Switzerland Famous pyramidal peak
JAPAN
Akita: Mineral Industry Museum
Tokyo: Natural Science Museum
* Fujiyama, Honshu Japan's highest (volcanic) peak
MALAYSIA
Ipoh: Geological Survey Museum
MEXICO
Mexico City: Museum of Geology
* Cacahuamilpa Caverns, Taxco Limestone caves
* Paricutín Fast-growing volcano that appeared in 1943
NAMIBIA
Lüderitz: Lüderitz Museum Diamond mining
NEPAL
* Mt. Everest The world's highest mountain
NETHERLANDS
Amsterdam: Geological Museum of the University of Amsterdam
Leiden: National Museum for Geology and Mineralogy
NEW ZEALAND
Dunedin: Geology Museum of the University of Otago
Wellington: New Zealand Oceanographic Institute
* Mt. Cook New Zealand's highest peak, much glaciated
NORWAY
Bergen: Geological Museum
Oslo: University Mineral-Geological Museum
* Jostedal Glacier Mainland Europe's largest icefield
* Sogne Fjord Norway's deepest, longest fjord
PERU
Cuzco: Geological Museum
* Colca Canyon The world's deepest canyon
POLAND
Warsaw: Museum of Earth Sciences
Wieliczka: Museum of the Cracow Salt Mines
PORTUGAL
Lisbon: Mineralogical and Geological Museum
ROMANIA
Ploesti: Museum of Petroleum
SAUDI ARABIA
Riyadh: Geological Museum
SOUTH AFRICA
Johannesburg: Bleloch Museum; Geological Museum
Kimberley: Kimberley Mine Museum
Pretoria: Museum of the Geological Survey
SPAIN
Barcelona: Geological Museum
Madrid: Museum of the School of Mines; National Geological Museum
* Cave of Nerja, Andalusia Has the longest stalactite
SUDAN
Khartoum: Geological Survey Museum

Aletsch Glacier, Switzerland

SWEDEN
Stockholm: Geological Research Museum
* **Stockholm Archipelago** Islands rising by isostatic rebound
SWITZERLAND
Zürich: Geological-Mineralogical Collection of the Confederation of Switzerland Technical College
* **Aletsch Glacier** Longest, largest glacier in the Alps
TANZANIA
* **Mt. Kilimanjaro** Africa's highest peak, a volcanic complex
THAILAND
Bangkok: Mineralogy Museum
TURKEY
Ankara: Turkish Natural History Museum
UNITED KINGDOM
Beamish, Durham: North of England Open-Air Museum
Blaunau Ffestiniog, Gwynedd: Gloddfa Ganol Slate Mine; Llechwedd Slate Caverns
Brandon, Norfolk: Grimes Graves Old flint mines
Castleton, Derbyshire: Blue-John Cavern and Mine
Clearwell, Gloucestershire: Clearwell Caves
Clitheroe, Lancashire: Clitheroe Castle Museum
Creetown, Dumfries and Galloway: Gem Rock Museum
Cynonville, West Glamorgan: Welsh Miners' Museum
Delabole, Cornwall: Delabole Slate Quarry and Museum
Dudley, West Midlands: Museum and Art Gallery
Glyn Ceiriog, Clwyd: Chwarel Wynne Mine and Museum
Llanberis, Gwynedd: The Welsh Slate Museum
London: Geological Museum The major British collection
Lound, Nottinghamshire: National Mining Museum
Manchester: Manchester Museum
Matlock Bath, Derbyshire: Peak District Mining Museum
Northwich, Cheshire: Salt Museum
Pendeen, Cornwall: Geevor Tin Mines
Ponterwyd, Dyfed: Llywernog Silver-Lead Mine
Rothesay, Isle of Bute: Bute Museum
St. Austell, Cornwall: Wheal Martyn Museum
Sandown: Museum of the Isle of Wight Geology
Sheffield, South Yorkshire: City Museum
Skipton, North Yorkshire: Craven Museum
Tenby, Dyfed: Tenby Museum
Wanlockhead, Dumfries and Galloway: Museum of Scottish Lead Mining
Wendron, Cornwall: Poldark Mine and Wendron Forge
* **Ben Nevis, Highland Region** Highest peak in the British Isles
* **Cheddar Caves and Gorge, Somerset** Limestone formations

Mt. Kilimanjaro, Tanzania

Cheddar Gorge, England

© DIAGRAM

245

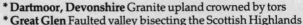

Geology displayed 3

Willamette Meteorite, Hayden Planetarium, New York City

Arches National Park, Utah

Carlsbad Caverns National Park, New Mexico

* **Dartmoor, Devonshire** Granite upland crowned by tors
* **Great Glen** Faulted valley bisecting the Scottish Highlands
* **White Cliffs of Dover, Kent** Chalk sea cliffs
UNITED STATES
Alma, Colorado: Alma Firehouse Museum Gold mining
Asheville, North Carolina: Colburn Memorial Mineral Museum
Aztec, New Mexico: Aztec Museum Association
Bisbee, Arizona: Bisbee Memorial Museum
Blacksburg, Virginia: Museum of the Geological Sciences
Boulder, Colorado: Pioneer Museum; University of Colorado Museum
Butte, Montana: Mineral Museum; World Museum of Mining
Calumet, Michigan: Coppertown USA
Carson City, Nevada: Nevada State Museum
Caspian, Michigan: Iron County Museum
Central City, Colorado: Central Gold Mine and Museum
Cincinnati, Ohio: Geology Museum
Coloma, California: Marshall Gold Discovery State Historic Park
Colorado Springs, Colorado: Western Museum of Mining
Columbia, Missouri: Geology Museum
Denver, Colorado: Denver Museum of Natural History
Des Moines, Iowa: Des Moines Center of Science and Industry
Eureka, Utah: Tintic Mining Museum
Laramie, Wyoming: University of Wyoming Geological Museum
Lexington, Kentucky: The Headley-Whitney Museum, Inc
Lincoln, Nebraska: University of Nebraska State Museum
Los Angeles, California: County Museum of Natural History
Louisville, Kentucky: Museum of Natural History and Science
Miami, Florida: Museum of Science
Nashville, Tennessee: Cumberland Museum and Science Center
New Haven, Connecticut: Peabody Museum Includes meteorites
New York City: The American Museum of Natural History; Hayden Planetarium Has a huge meteorite
Nome, Alaska: Carrie McLain Museum Gold-mining exhibits
Norman, Oklahoma: Stovall Museum of Science and History
Philadelphia, Pennsylvania: Wagner Free Institute of Science
Phoenix, Arizona: Arizona Mineral Museum
Pittsburgh, Pennsylvania: Carnegie Museum of Natural History
Portales, New Mexico: Miles Museum
Portland, Oregon: Oregon Museum of Science and Industry
Providence, Rhode Island: Roger Williams Park Museum
Rapid City, South Dakota: Museum of Geology
Richmond, Virginia: Science Museum of Virginia
Rochester, New York: Rochester Museum and Science Center
Sacramento, California, Science Center and Junior Museum
St. Paul, Minnesota: The Science Museum of Minnesota
Salt Lake City, Utah: Hansen Planetarium Has meteorites
San Francisco, California Division of Mines and Geology
San Jose, California: Rosicrucian Planetarium and Science Museum
Seattle, Washington: Pacific Science Center
Springfield, Massachusetts: Springfield Science Museum
Tucson, Arizona: University of Arizona Mineralogical Museum
Washington, D.C., National Museum of Natural History
* **Arches National Park, Utah** Natural sandstone arches
* **Badlands, South Dakota** Dissected, barren upland
* **Bryce Canyon, Utah** Eroded maze of rock pinnacles

* **Carlsbad Caverns National Park, New Mexico**
* **Crater Lake, Oregon** Lake in a caldera
* **Devils Tower, Wyoming** Upstanding, columnar volcanic rock
* **Grand Canyon, Arizona** The world's largest land gorge with rocks recording many million years of Earth history
* **Great Salt Lake, Utah** The Americas' largest salt lake
* **Cape Hatteras, North Carolina** Part of a 200mi (320km) chain of sandspits
* **Malaspina Glacier, Alaska** Largest US glacier
* **Mammoth Cave, Kentucky** Longest of all known cave systems
* **Mauna Loa, Hawaii** Earth's largest active volcano
* **Mt. McKinley, Alaska** North America's highest peak
* **Meteor Crater, Arizona** Famous meteorite crater
* **Monument Valley, Arizona/Utah** Sandstone buttes
* **Niagara Falls, New York** Famous waterfall
* **Rainbow Bridge, Utah** The world's largest natural bridge
* **Mt. Rainier, Washington** Highest peak in the volcanically active Cascade Range
* **San Andreas Fault, California** Earthquake-prone fault running from north of San Francisco to the Gulf of California
* **Stone Mountain, Georgia** Granite dome
* **White Sands National Monument, New Mexico** Includes part of the world's largest gypsum dune field
* **Yellowstone National Park, Wyoming/Montana/Idaho** Site of some 300 geysers and 10,000 hot springs
* **Yosemite National Park, California** Includes vast glaciated granite domes and spectacular waterfalls
USSR
Leningrad: Central Geological and Prospecting Museum
Moscow: A.E. Fersman Mineralogical Museum
Moscow: Museum of Earth Science
* **Lake Baikal, Siberia** The world's deepest lake and largest freshwater body, lying in an immense crustal trough
VENEZUELA
Caracas: Museum of Natural Sciences
* **Angel Falls** The world's highest waterfall
YUGOSLAVIA
Belgrade: Nature Museum
ZAIRE
* **Boyoma Falls** The world's largest falls by volume
ZAMBIA/ZIMBABWE
* **Victoria Falls** Africa's most dramatic waterfall
ZIMBABWE
Harare: Macgregor Museum

Devils Tower, Wyoming

Hot spring terraces,
Yellowstone National Park

Lake Baikal, USSR

© DIAGRAM

247

FURTHER READING

General

Calder, Nigel *Restless Earth* British Broadcasting Corporation, 1972
Clark, I.F. and Cook, B.J., eds. *Perspectives of the Earth* Australian Academy of Science, 1983
Garland, George D. *Introduction to Geophysics* W.B. Saunders Co, 1979
Institute of Geological Sciences *The Story of the Earth* HMSO, 1972
Matthews, W.H. *Geology Made Simple* W.H. Allen, 1970
Mitchell, J., ed. *The Joy of Knowledge Encyclopaedia* Guild Publishing, 1980
Potter, A.W.R. and Robinson, H. *Geology* Macdonald and Evans, 1982
Smith, D.G., ed. *Cambridge Encyclopedia of Earth Sciences* Cambridge University Press, 1982
Smith, J., ed. *Dictionary of Geography* Charles Letts, 1983
Note: Recent major discoveries appear in quality newspapers, and in general and specialist science
  magazines.

Chapter 1: Sizing up the Earth

Kandel, R.S. *Earth and Cosmos* Pergamon Press, 1980
Ozima, M. *The Earth its birth and growth* Cambridge University Press, 1981
Perrin, M.B. *An Introduction to the Chemistry of Rocks and Minerals* Edward Arnold, 1979

Chapter 2: The Restless Crust

Clark, I.F. and Cook, B.J. eds.: see above.
Gass, I.G., Smith, P.J., and Wilson, R.C.L. *Understanding the Earth* Artemis Press, 1972
Smith, D.G., ed.: see above.
Weyman, D. *Tectonic Processes* Allen & Unwin, 1981
Windley, B.F. *The Evolving Continents* Wiley, 1984

Chapter 3: Fiery Rocks

Clark, I.F. and Cook, B.J., eds.: see above.
Smith, D.G., ed.: see above.
Van Rose, S. and Mercer, I *Volcanoes* HMSO, 1974

Chapter 4: Rocks from Scraps

Clark, I.F. and Cook, B.J.: see above.
Pettijohn, F.J. *Sedimentary Rocks* Harper & Row, 1975

Chapter 5: Deformed and Altered Rocks

Clark, I.F. and Cook, eds.: see above.
Smith, D.G., ed.: see above.

Chapter 6: Crumbling Rocks through **Chapter 9: The Work of Ice and Air**

Bryant, R.H. *Physical Geography Made Simple* Heinemann, 1979
Bunnett, R.B. *Physical Geography in Diagrams* Longman, 1976
Clowes, A. and Comfort, P. *Process and Landform* Oliver & Boyd, 1982
Holmes, A. *Principles of Physical Geography* Thomas Nelson & Sons Ltd, 1969
Monkhouse, F.J. *Principles of Physical Geography* University of London Press, 1971
Strahler, A.N. and Strahler, A.H. *Modern Physical Geography* Wiley, 1987
Tarbuck, E.J. and Lutgens, F.K. *Earth Science* Merrill, 1979

Chapter 10: Change Through the Ages and **Chapter 11: The Last 600 Million Years**

Dott, R.H., Jr. and Batten, R.L. *Evolution of the Earth* McGraw-Hill, 1976
Redfern, R. *The Making of a Continent* British Broadcasting Corporation, 1983
Seyfert, C.K. and Sirkin, L.A. *Earth History and Plate Tectonics* Harper & Row, 1979
Van Andel, Tjeerd H. *New Views on an Old Planet* Cambridge University Press, 1985
Windley, B.F. *The Evolving Continents* Wiley, 1984

Chapter 12: Rocks and Man

Barnes, J.W. *Basic Geological Mapping* Open University Press, 1981
Cargo, D.N. and Mallory, B.F. *Man and His Geologic Environment* Addison-Wesley, 1977
Reader's Digest Book of Natural Wonders Reader's Digest, 1980
Firsoff, V.A. and Firsoff, G.I. *The Rockhound's Handbook* David and Charles, 1975
Kouřimský, J. *The Illustrated Encyclopedia of Minerals and Rocks* Octopus Books, 1977
Richey, J.E. *Elements of Engineering Geology* Pitman, 1964
Wood, R.M. *The Dark Side of the Earth* Allen & Unwin, 1985

INDEX